Lecture Notes in Economics and Mathematical Systems

562

Founding Editors:

M. Beckmann
H. P. Künzi

Managing Editors:

Prof. Dr. G. Fandel
Fachbereich Wirtschaftswissenschaften
Fernuniversität Hagen
Feithstr. 140/AVZ II, 58084 Hagen, Germany

Prof. Dr. W. Trockel
Institut für Mathematische Wirtschaftsforschung (IMW)
Universität Bielefeld
Universitätsstr. 25, 33615 Bielefeld, Germany

Editorial Board:

A. Basile, A. Drexl, H. Dawid, K. Inderfurth, W. Kürsten, U. Schittko

Lecture Notes In Economics
and Mathematical Systems

Tobias Langenberg

Standardization
and Expectations

 Springer

Author

Tobias Langenberg
University of Hamburg
Department of Economics
Institute of Public Finance
Von-Melle-Park 5
20146 Hamburg
Germany

ISSN 0075-8442
ISBN-10 3-540-28112-6 Springer Berlin Heidelberg New York
ISBN-13 978-3-540-28112-2 Springer Berlin Heidelberg New York

This work is subject to copyright. All rights are reserved, whether the whole or part
of the material is concerned, specifically the rights of translation, reprinting, re-use of
illustrations, recitation, broadcasting, reproduction on microfilms or in any other way,
and storage in data banks. Duplication of this publication or parts thereof is permitted
only under the provisions of the German Copyright Law of September 9, 1965, in its
current version, and permission for use must always be obtained from Springer-Verlag.
Violations are liable for prosecution under the German Copyright Law.

Springer is a part of Springer Science+Business Media

springeronline.com

© Springer-Verlag Berlin Heidelberg 2006
Printed in Germany

The use of general descriptive names, registered names, trademarks, etc. in this
publication does not imply, even in the absence of a specific statement, that such
names are exempt from the relevant protective laws and regulations and therefore
free for general use.

Typesetting: Camera ready by author
Cover design: *Erich Kirchner*, Heidelberg

Printed on acid-free paper   42/3130Jö   5 4 3 2 1 0

Preface

This book has been accepted as my doctoral thesis by the Department of Economics at the University of Hamburg. First of all, I am indebted to my supervisor, Prof. Manfred Holler, for his scientific guidance and support. At his Institute of SocioEconomics, I have found an environment that was ideal for my research. I would also like to thank Prof. Hans-Joachim Hofmann and Prof. Wilhelm Pfähler for helpful comments.

I am also very grateful to Prof. Gunther Engelhardt for supporting me during my time as Teaching and Research Assistant at the Institute of Public Finance, University of Hamburg.

I have received valuable comments from a large number of people. I wish to thank in particular my colleagues Heide Coenen, Jörg Gröndahl and Ingolf Meyer Larsen for interesting and inspiring discussions.

Finally, I thank my parents and my brother for their encouragement and help.

Hamburg, September 2005   *Tobias Langenberg*

Contents

1

Introduction

Over the last decades, technological progress has brought about a multitude of standardization problems. For instance, compatibility standards ensure the interoperability of goods, which is of decisive importance when users face positive externalities in consumption. These so-called "network externalities" refer to goods such as telephones or fax machines, which would generate only small benefits if they were adopted by few users. Such communication networks involve direct network effects in that the consumption benefit of a single user directly increases with the number of network participants. The existence of network externalities suggests that the allocation in network markets may be inefficient. Typically, the buyer of a network good takes into account his private costs and benefits without internalizing the network benefits he generates for other network participants. Thus, an important question in the economics of network effects and standardization is whether network markets bring about efficient standards.

Since the early contributions by David (1985), Farrell and Saloner (1985), Katz and Shapiro (1985), a vast literature on network effects and standardization has been evolving. But yet, little attention has been devoted to the formal analysis of how standardization and consumers' expectations interact. Expectations are of decisive importance of whether a new technology will prevail as de-facto standard or not. Early adopters must be confident that the network good will be successful. Thus, it may be worthwhile for firms to influence expectations. A classical tactic aimed at influencing expectations is product pre-announcement. By pre-announcing its upcoming technology, a firm may increase the expected network size of its new technology to the disadvantage of the rival's technology. For instance, in the mid 1980s, Borland released its new spreadsheet Quatro Pro. However, its main rival Microsoft thwarted Quatro Pro's growth by pre-announcing (and praising) the next release of its competing software, Excel.[1]

Economides (1996a) discusses an alternative way to influence expectations. He shows that it can be worthwhile for an incumbent monopolist to share its technology with competitors. What drives this model result is the assumption that high expected sales increase consumers' willingness to pay for the network good. By inviting competitors into its network, the incumbent firm can credibly commit to a network size which exceeds its profit-maximizing monopoly quantity. Thus, the incumbent firm faces a tradeoff. On the one hand, the invitation of rival firms increases the equilibrium network size and thus consumers' willingness to pay via network effects. On the other hand, the invitation of rival firms involves competition. For a given level of expected sales, this "competition effect" has a negative impact on the incumbent's profit.

This type of expectation management can also be applied to the case of indirect network effects and systems competition.[2] Then, the supplier of a hardware-software system may invite independent suppliers of compatible software products, thereby committing credibly to a large variety of software. Alternatively, buyers would run the risk of facing a small variety of software in the future. Due to high switching costs, they might be "locked-in" to the corresponding hardware-software system.[3] IBM's strategy of licensing its technology to independent hardware and software manufacturers gives an example for successful expectation management to establish the PC standard. The rival Apple-Mac network followed another strategy. The first ten years after the introduction of the Mac, Apple refused to license independent manufacturers, so-called clones. As a consequence, Apple's market share constantly decreased.

Thus, numerous examples suggest that expectations are "a key factor in consumer decisions about whether or not to purchase a new technology,..."

---

[1] See Farrell and Saloner (1986a) for a formal analysis of product pre-announcements.
[2] See Holler, Knieps and Niskanen (1997) for an overview of various models with network effects.
[3] See Klemperer (1987), Farrell and Shapiro (1988), Arthur (1989), Beggs and Klemperer (1992) and Witt (1997) for a discussion of consumer lock-in.

(Shapiro and Varian, 1999, p. 275). Consisting of three essays on various aspects of standardization and expectations[4], this thesis aims at deepening our understanding of how standards and expectations interact. The analysis puts an emphasis on the following main questions:

1. How may existing standards affect the agents' expectations?
2. How may expectations affect the evolution of standards?
3. What are the welfare implications of the equilibrium, and which solutions would be imposed by a "social planner"?

The main purpose of the first essay is to find economic reasons why university examinations should be standardized, *i.e.* why the requirements should be comparable among different universities. The essay refers to the main question of how standards may affect agents' (*i.e.* employees' and employers') expectations. Here, standardization is considered as a means of reducing variation in examination requirements. This kind of reference standard may be realized by introducing central examinations. Or alternatively, diplomas should qualify for accreditation by certification bodies.

Starting from the basic signaling model, taken from Spence (1973), the first essay analyzes the welfare implications of signaling. Whereas signaling is only a distributive device in the basic model, an extension of the model shows that signaling may increase total output by enabling correct matching of employees to jobs. If examination requirements vary among universities, the job-matching effect deteriorates. This situation of incomplete information about the signal's quality is formalized as a Bayesian Game. Employers and employees are assumed to know the distribution of examination requirements. On the basis of this common knowledge, employers form expectations about whether a signaling employee belongs to the more productive type or not. By standardizing the requirements, the educational signal regains reliability and recovers its job-matching function. However, there is a tradeoff between the job-matching function and total signaling costs.

The second essay analyzes the competition between two firms when their incompatible technologies exhibit network effects. We mainly refer to the

---

[4]   The analysis is not confined to compatibility standards and network effects. In fact, it also deals with so-called reference standards, which facilitate the transaction of complex goods by describing product features. See 2.1, for a taxonomy of standards.

problem of how compatibility standards may affect consumers' expectations.[5] Our framework distinguishes between two different regimes of standardization. Whereas the first regime involves that firms compete within a joint network (*intra-technology competition*), the second regime refers to standardization by means of blockaded or deterred entry of a rival technology (*inter-technology competition*).

Following Economides (1996a), we assume that high expected sales increase the willingness to pay for the corresponding good. At the equilibrium level, consumers' expectations have to be fulfilled. Whereas the model by Economides is confined to intra-technology competition, we will analyze both intra-technology and inter-technology competition. An incumbent firm faces the strategic choice of whether to share its superior technology (via free licensing) with a follower or to keep its technology for itself. The first option of sponsoring intra-technology competition increases the incumbent firm's network and thus consumers' willingness to pay because the incumbent credibly commits to a larger network. On the other hand, the latter option involves inter-technology competition. Depending on the relative cost advantage of the incumbent firm, the entry of the rival technology may be blockaded, both technologies can coexist in an incompatible duopoly or the incumbent firm may deter the market entry of its rival. The essay investigates the incumbent firm's choice of whether to sponsor intra-technology competition or to insist on inter-technology competition.

The third essay deals with standardization of nascent technologies. A common characteristic of nascent technologies is that consumers cannot completely assess the product's quality at the time of market launch. We make the assumption that consumers learn about the actual stand-alone value of a technology after using it ("learning by using"). Before using the technology, consumers are assumed to know the distribution of stand-alone values, only. We will present a two-period framework with two competing network technologies and two consumers. In the first period, consumers may adopt incompatible technologies (*experimentation*), or they can choose a joint technology (*ex-ante standardization*). In the second period, the stand-alone values of all tech-

---

[5] However, the second essay also touches on the subject of how expectations may affect the evolution of standards. The incumbent's strategic choice between inter-technology and intra-technology competition involves multiple equilibria. Thus, consumers' expectations determine the evolution of standards.

nologies used in the first period become public knowledge. Based on this information, each user chooses among three options: Firstly, the user may stick to his technology. As a second option, he can switch to the other technology. Finally, the user may choose an "outside option".

Ex-ante standardization is related to consumers' expectations inasmuch as it involves *limited information* in the second period: Consumers only find out the actual stand-alone value of the joint technology which they have chosen as ex-ante standard. On the basis of the observed stand-alone value, consumers form expectations about the alternative technology (which they have not yet used), *i.e.* they revise the expected ex-ante value according to the Bayesian rule. Experimentation allows consumers to find out the actual stand-alone values of *both* technologies so that their choice of the ex-post standard is based on *complete information*. However, experimentation involves a transient or even persistent loss of compatibility. By means of ex-ante standardization, consumers enjoy network benefits from the beginning. Thus, there exists a tradeoff between ex-ante standardization and experimentation.

The third essay also refers to the second main question of how consumers' expectations affect the evolution of standards. Consumers' ex-ante expectations about the technologies' values are represented by the joint probability distribution, which is "common knowledge". We will analyze the impact of different parameters such as correlation, variance and expected values on the equilibrium values. For the sake of traceable results, we will assume that the values of two potential technologies are drawn from a bivariate normal distribution. The numerical analysis demonstrates that consumers prefer ex-ante standardization to experimentation if they expect the values of both technologies to be strongly correlated. Furthermore, the model shows that if the technologies are not equally attractive ex ante, there can be too much ex-ante standardization compared with the social optimum, or consumers may choose an inferior technology as ex-ante standard.

Table 1.1 shows a classification of the three essays with respect to the considered type of standard and the problem of how standards and expectations interact. Since each essay deals with the third main question of how a social planner should intervene, this problem is omitted in our classification. The structure of this thesis arises from the classification. Chapter 2 is devoted to a brief introduction to the concept of network effects and standardization, which is relevant for the understanding of subsequent chapters. Chapter 3 contains the first essay dealing with standardization of educational signals and job matching. The first essay is related to the category of reference standards. In Chapter 4, we will present the second essay which is devoted to the problem of inter-technology versus intra-technology competition in network markets. Chapter 5 contains the third essay on standardization of nascent technologies. Finally, Chapter 6 briefly summarizes the main results of this thesis.

Table 1.1: Classification of models

Type of Standard	How does standardization affect agents' expectations?	How do agents' expectations affect the evolution of standards?
Reference Standards	**Model I:** Standardization as a means of reducing variation in examination requirements. Tradeoff between job-matching effect and total signaling costs.	Standardization process is exogenous.
Compatibility Standards	**Model II:** Standardization as a credible commitment to a larger network. Standardization by sponsoring intra-technology competition or as the result of inter-technology competition. Tradeoff between network size and competition.	Incumbent's strategic choice between intra-technology and inter-technology competition involves multiple equilibria. Consumers' expectations determine evolution of standards.
	Model III: Standardization of nascent technologies. Experimentation involves complete information. Ex-ante standardization leads to limited information. Choice of the ex-post standard is based on expectations, i.e. users form expectations about the rival technology's value according to the Bayesian Rule. Tradeoff between early standardization and experimentation.	Parameters of the joint probability distribution ("common knowledge") determine evolution of standards.

2

The Economics of Standardization: Basic Concepts

This chapter gives a brief introduction to the economics of standardization, which we refer to in subsequent chapters. Section 2.1 deals with a taxonomy of standards. In Section 2.2, we will discuss the concepts of network effects and compatibility standards used in this thesis.

2.1 Taxonomy of Standards

Generally speaking, standardization includes doing certain key things in a uniform way.[6] Standardization may occur in a multitude of forms. For example, the term "standardization" can refer to labeling standards fixing a maximum for the proportion of chicken allowable in a "beef frankfurter" or to the physical design of computer interfaces ensuring compatibility to printers.

Following David (1987), who provides a useful taxonomy, we differentiate between *technical standards* and *standards for human behavior*. Whereas technical standards refer to features of an inanimate object (*i.e.* its material and design properties), the latter category applies to human behavior, procedures and performance. Obviously, technical standards are easier to specify in a quantitative manner than its behavioral counterparts.

As Table 2.1 depicts, technical and behavioral standards may take three different forms. For example, technical *reference standards* describe a reference point such as currencies, weights or measures of materials and products. Behavioral reference standards are exemplified by precedents in law and accreditation of institutions.

Technical standards *for minimal admissible attributes* define a cardinal minimum bound, as exemplified by safety levels or standards for product

---

[6] See Farrell and Saloner (1992), p. 9.

quality.[7] The behavioral analogs of this category are, for instance, job qualifications[8], which define a minimal level of educational attainment (minimal scores in exams) or legal codes, which separate legal from illegal conduct.[9]

Table 2.1: Taxonomy of Standards, following David (1987)

	Standards of Technical Design	Standards of Behavioral Performance
Standards for Reference	currencies, weights, measures of materials or products	precedents in law, accreditation of institutions
Standards for Minimal Admissible Attributes	safety levels, product quality	legal codes, job qualifications
Standards for Interface Compatibility	physical design of interfaces	language standards, standards of commercial conduct ("honesty")

Technical *compatibility standards* ensure the interoperability of goods, which is of decisive importance in the case of network goods such as telephones, fax machines and computers. The behavioral counterparts of physical networks are "metaphorical networks" (Liebowitz and Margolis, 1994, p. 136), such as the network of English speakers, which allows for interrelationship among

---

[7] See Jones and Hudson (1996 and 1997) for an economic analysis of quality standards. By reducing the variance of product quality, standardization reduces the costs of search in this approach. The pioneering paper to analyze problems of asymmetric information is Akerlof (1970).

[8] However, job qualifications could also refer to the category of compatibility standards, as pointed out by Layes (1998).

[9] For a discussion of law as a standardizing system, see Adams (1996) and Jørgensen (1997).

speakers without having physical connections.[10] Compatibility standards may also be beneficial without explicitly assuming network externalities. Matutes and Regibeau (1988) and Economides (1989) study in mix-and-match models the effects of compatibility when consumers can buy "hybrid systems", which are composed of vertically compatible components (such as tuner, CD player and loudspeakers) from different manufacturers.

As David (1987) points out, it is difficult to categorize standards according to their ultimate economic effects because a given standard can perform different economic functions. For instance, a telecommunication standard such as UMTS has a compatibility component which gives rise to network effects. Furthermore, technical specifications may result in a reduction of variety, thereby reducing transaction costs, *i.e.* problems of information asymmetries between firms and consumers may be reduced.[11]

Standards may arise in a number of ways. Market-mediated or *de-facto standards* are determined by market forces. De-facto standardization may occur in markets with sponsored or non-sponsored technologies. The class of sponsored technologies involves that each competing technology is held by a small number of firms. Then, firms may use instruments such as "penetration pricing" or "product pre-announcements" to establish their proprietary technology as de-facto standard. Penetration pricing refers to the technique of offering low prices to early customers in order to build up an installed base and to influence the choices of later user. By means of product pre-announcements, a network sponsor tries to retard the growth of its rivals' networks.[12]

The ideal type of non-sponsored technologies arises from the situation where rival technologies are each supplied by a large number of firms. If products are completely homogeneous with respect to each technology, firms

---

[10] See, for instance, Church and King (1993) who develop a model in which the benefit of language acquisition is increasing in the number of individuals who speak the language. This gives rise to network externalities and, if language acquisition is costly, individuals may come to inefficient decisions.

[11] Farrell and Saloner (1986b) demonstrate that there is a tradeoff between standardization and variety. If consumers have heterogeneous preferences with respect to the good specifications, standardization reduces the matching with preferences.

[12] For strategies and tactics in de-facto standardization, see Besen and Farrell (1994), Katz and Shapiro (1994) and Belleflamme (2002).

set uniform prices equal to the marginal costs of the corresponding technology, *i.e.* a single firm has no market power in this framework.[13]

As an alternative to the market mechanism, standards may be enforced by the government. These so-called *de-jure standards* can be classified into two ideal types, the bureaucracy and the committee solution. The first type refers to the situation where standards are formulated and enforced by governmental agencies. In the case of the committee solution, the involved parties negotiate over the standard and the negotiated standard is enforced by the government. These committees may consist of standardization bodies and of stakeholders such as single firms, consumer and industry organizations. Alternatively, committee standards may be based on voluntary cooperation, *i.e.* they are non-enforced by the government.[14]

After having presented different categories of standards, we are able to specify the standards, which we refer to in this thesis. In the first essay (Chapter 3), we examine reference standards which reduce variation in examination requirements. These *de-jure* standards may be formulated and imposed by governmental agencies or by committees comprising stakeholders and representatives of accreditation bodies. The second essay (Chapter 4) deals with de-facto standardization and sponsored network technologies, *i.e.* each technology is held by a single firm and standards are determined by market forces. The third essay (Chapter 5) is based on the implicit assumption that technologies are non-sponsored. We will analyze the coordination problems of standardization (*i.e.* market vs. committee solution) from the perspective of consumers.

---

[13] See Thum (1995) for the distinction between sponsored and non-sponsored technologies.

[14] Thum (1994) compares the welfare implications of the bureaucracy and the committee solution.

2.2 Compatibility Standards and the Concept of Network Effects

2.2.1 Direct Network Effects

Classic examples of goods which give rise to direct network effects are communication goods such as telephones and fax-machines.[15] Using such goods not only confers a benefit on the consumer himself but also on the other consumers using the same or compatible goods. Compatibility can also be a matter of degree rather than being a matter of "all or nothing" (see, for instance, Blankart and Knieps, 1993). The larger the number of network participants, the larger the value to being part of this communication network. Typically, communication goods such as telephones have a stand-alone value equal to zero, *i.e.* in the absence of potential communication partners, these goods confer no benefit on a user.[16]

In the literature, direct network effects are formalized in different ways.[17] The same is true for the network approaches we will present in this thesis. In the third essay (Chapter 5), users simply add a constant network value n to the stand-alone value of their technologies if they use identical technologies. However, in the second essay (Chapter 4), direct network effects rely on the expected number of compatible units to be sold. The model assumes rational consumers, *i.e.* in the fulfilled expectations equilibrium, actual sales must be equal to expected sales. Fig. 2.1 shows the construction of a fulfilled expectations demand curve. The curves $p(y, y_i^e)$ depict consumers' willingness to pay for a varying quantity y, given the expected sales y_i^e. At $y = y_i^e$, expectations are fulfilled and the point belongs to the fulfilled expectations demand $p(y, y)$. Thus, $p(y, y)$ can be constructed by connecting all the points with identical expected and real sales.

---

[15] For an analysis of communication goods, see Blankart and Knieps (1994).

[16] For a classification of networks, see Economides (1996b).

[17] See the pioneering papers by Katz and Shapiro (1985) and by Farrell and Saloner (1986a, 1986b). For later works on direct network effects, see, for instance, de Palma and Leruth (1996), Economides (1996a), Chou and Shy (1996), and Jeanneret and Verdier (1996).

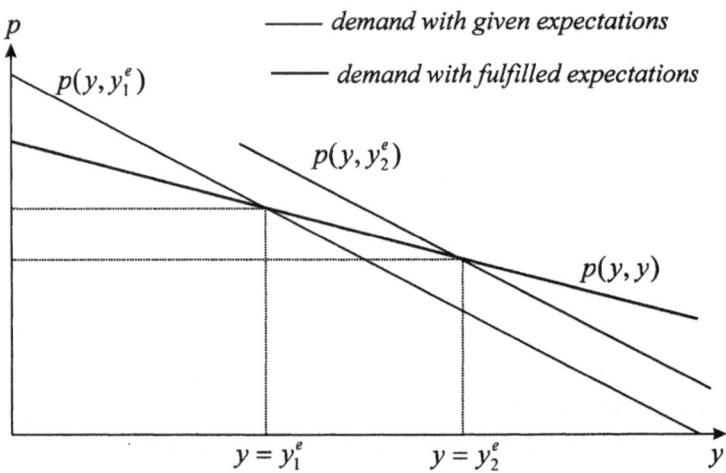

Fig. 2.1: Construction of a fulfilled expectations demand curve

2.2.2   Indirect Network Effects

Whereas direct network effects refer to the fit of functionally equivalent goods (horizontal compatibility), indirect network effects may occur in markets with systems. Systems consist of perfectly complementary products such as hardware and software. In the case of hardware-software systems, a hardware user's benefit rises with the number of complementary software products being vertically compatible to the corresponding hardware.[18] For instance, the buyers of video cassette recorders (hardware) attach great importance to a manifold supply of movies and shows which are available on compatible video cassettes.[19] Consequently, extending the number of compatible software products can increase the attractiveness of a hardware-software system. Church and Gandal (1996) assume that a hardware supplier can directly increase the installed base of complementary software by vertical integration. Chou and Shy (1990) model an indirect mechanism of increasing the installed base of software. In their framework, a hardware user benefits from a large

---

[18] See Wiese (1997) for the distinction between horizontal and vertical compatibility.

[19] For empirical studies on indirect network effects, see, for instance, Saloner and Shepard (1995), Koski (1999), and Gandal, Kende and Rob (2000).

number of users buying compatible hardware, not because of a larger number of potential communication partners as in the case of communication goods, but due to the fact that additional hardware users make for a larger variety in the market for compatible software. Typically, the production of a particular software program is only possible if demand exceeds a threshold. If software markets are characterized by monopolistic competition and software firms face economies of scale, then the variety of software increases with the demand of compatible hardware.[20]

---

[20] See, for instance, Chou and Shy (1990) and Church and Gandal (1992, 1993) for pioneering papers on indirect network effects.

3

Standardization of Educational Signals and Job Matching

The purpose of this chapter is to find economic reasons why examination requirements should be comparable among different universities. Starting from the basic signaling model, taken from Spence (1973), we will analyze the welfare implications of signaling. Whereas signaling is only a distributive device in the basic model, an extension of the model shows that signaling may increase total output by enabling a correct matching of employees to jobs. If examination requirements vary among universities, the job-matching effect deteriorates. This situation of incomplete information about the signal's quality is formalized as a Bayesian Game. By standardizing the requirements, for example by means of central examinations, signaling recovers its job-matching function. However, there is a tradeoff between the job-matching function and total signaling costs.

3.1   Introduction

Educational standards have been the subject of public and scientific interest, but the formal analysis of this problem lags behind. What are the economic reasons why examination requirements should be comparable among different universities? One possible explanation is due to the signaling theory, which considers education as a device to reveal the candidate's ability.

The basic signaling model, which goes back to Spence (1973), relies on the assumption that education does not enhance the graduates' productivity.[21] Nevertheless, it enables them to signal their innate productivity to the employers. The point of departure is an information asymmetry between employers

---

[21]   See Spence (2002) for a retrospect on signaling models.

and job applicants. While the job applicants know their true productivity, the employers are not able to observe it directly. For productive students, signaling is a device to distinguish themselves from less productive job applicants in order to receive a higher wage. The crucial assumption in the signaling approach is that mental and time costs of the signal acquisition are negatively correlated with the given productivity of the students. Thus, in a signaling-separating equilibrium, only students with strong capabilities are able to acquire the signal.

In an extension of this basic signaling model, we assume that the productivity of a worker not only depends on the worker type but also on the type of employer. The analysis suggests that signaling may increase total output by enabling a correct matching of employees to jobs. There are already models dealing with the impact of educational signals on job matching. Belman and Heywood (1997) assume that the employees acquire imperfect educational signals. In the first period, employers cannot observe the true productivity of the workers. Because of the signal's imperfection, the matching process results in non-optimal matches. In the second (and last) period, the workers' productivity is revealed and the mismatched workers are reassigned to appropriate jobs. Hence, the average productivity of all workers improves from period one to period two. The main result is that the return to an educational signal declines over time. Like Belman and Heywood, we assume that the productivity of a worker in a particular job depends on the quality of the match between the requirements of the job and the innate ability of a worker. However, in contrast to their approach, we explicitly model the *Spencian* process of signal acquisition that depends on the relative costs and benefits of each worker type.[22] It is shown that, given the self-selection conditions hold, a signaling-separating equilibrium (*SSE*) and a non-signaling-pooling equilibrium (*NSPE*) may exist. If the more productive employee type faces relatively low signaling costs, the *SSE* is unique. For relatively high signaling costs, both the *SSE* and the *NSPE* occur, *i.e.* there are multiple equilibria.[23]

---

[22] See Langenberg (2002) for a preliminary version of the job-matching model.

[23] See Holler, Layes and Winckler (1999) for a similar framework with high- and low-quality workers who are organized in respective unions. The model investigates the effects of downward compatibility, *i.e.* high-quality workers are allowed to work on low-quality jobs which are otherwise reserved for low-quality workers.

The welfare implications of job matching are analyzed by means of a simple additive welfare function. The model shows that signaling may increase welfare by enabling a correct matching of employees to jobs. But the *SSE* can be inefficient: From the welfare perspective, signaling is not desirable if signaling costs outweigh positive job-matching effects. However, the more productive employees also internalize the welfare-neutral distributive effect of signaling. For them, signaling is a device to claim a higher share of the total output because they avoid subsidizing the less productive type.

Moreover, the model deals with the problem of signal imperfection. Whereas Belman and Heywood merely assume that the signal is imperfect, we consider a range of different examination requirements. This situation of incomplete information about the signal's quality is formalized as a Bayesian Game and the equilibria and welfare implications are analyzed numerically. The signaling function of education is less effective if universities have different examination requirements. By standardizing the requirements, for example by means of central examinations, signaling can recover its filtering function. However, there is a tradeoff between the job-matching function and total signaling costs.[24]

The structure of this chapter is as follows. In Section 3.2, a modified version of the basic signaling model is presented. Section 3.3 deals with the job-matching effect of educational signals. In Section 3.4, we consider a continuum of examination requirements in order to study the effects of incomplete information. Equilibria and welfare implications are analyzed numerically. In Section 3.5, we derive equilibria in mixed strategies for the case of perfect educational signals. This analysis is based upon the framework of Section 3.4. Concluding remarks follow in Section 3.6.

3.2   The Basic Signaling Model

Assume that there are two different worker types: Let b denote the productivity of type B, which is assumed to be higher than a, the productivity of type A. Thus, the productivity advantage of type B over type A is given by $\Delta = b - a > 0$. Let q denote the proportion of type A workers. Whereas each

---

[24] For a standardization model based on the human-capital theory, see Costrell (1994)

worker knows his own type, the employers cannot directly observe the true productivity of a worker. But the employers are able to measure the total output, ex post. It follows from profit maximization that the employers pay a wage according to the expected productivity of an employee, *i.e.* they form expectations about the type of each worker.

Each employee is able to invest in higher education in order to signal his given productivity. The crucial assumption is that type B incurs lower costs of signal acquisition than the less productive type A. Let the signaling cost of type A be equal to $C_A = y$, whereas type B bears a cost of $C_B = \mu y$ with $0 < \mu < 1$. Here, y denotes the level of examination requirements.

A signaling-separating equilibrium (*SSE*), given by $y^* > 0$, has the following characteristics: If a worker has acquired the signal, his employer expects him to be of type B (with a probability of one) and pays him a wage equal to b. Employees without signal are considered to be less productive and get a wage equal to a. In equilibrium, employers' expectations must be fulfilled, *i.e.* there must exist a level of requirements y^* that induces all type B employees to decide voluntarily in favor of signaling, whereas all type A employees voluntarily choose the non-signaling strategy. This problem can be formalized by the following self-selection conditions:

$$a > b - y^*,\tag{3.1}$$

$$b - \mu y^* > a.\tag{3.2}$$

Both inequations can be combined to

$$\Delta < y^* < \Delta/\mu.\tag{3.3}$$

In order to examine for which parameters the *SSE* is unique, assume that y^* is marginally higher than the minimum value Δ. Suppose that all employees decide against signaling. If no employee acquires the signal, the employers are unable to distinguish both worker types. Then, the employers pay a uniform wage which is equal to the average productivity of all employees. In this case, type B "subsidizes" type A:

$$\overline{w} = q a + (1 - q)b = b - \Delta q.\tag{3.4}$$

The more type B employees decide against signaling, the higher is the non-signaling payoff. Thus, (3.4) shows the maximal non-signaling payoff because no type B employee chooses signaling. If only one type B employee decided against signaling, the non-signaling payoff would equal a.

The non-signaling-pooling payoff, given in (3.4), must be compared with the payoff in the signaling-separating case. The *SSE* is unique if the following condition holds:

$$b - \mu\Delta > b - \Delta q, \qquad (3.5)$$

$$\Rightarrow \quad q > \mu.$$

A single type B employee would always choose the signaling strategy even if all remaining type B employees decided against signaling because the signaling payoff $b - \mu\Delta$ exceeds the non-signaling payoff $b - \Delta q$. Since the same is true for all type B members, the *NSPE* is not feasible in this situation.

If $\mu > q$, multiple equilibria occur. Suppose that all type B members decide against signaling. In this case, a single type B employee has no incentive to deviate from the non-signaling strategy because the non-signaling payoff $b - \Delta q$ exceeds the signaling payoff $b - \mu\Delta$, *i.e.* the *NSPE* exists. Nevertheless, the *SSE* is feasible as well: Assume that all type B employees decide in favor of signaling. A single type B employee has no incentive to deviate from the signaling strategy because the signaling payoff $b - \mu\Delta$ is higher than the non-signaling payoff, which is equal to a in this situation. Note that the *SSE* can be interpreted as a coordination failure in the case of multiple equilibria because the *NSPE* would make each type B employee better off.

After having discussed the equilibria, we can show that signaling has negative welfare implications. In contrast to the human-capital approach, education has no impact on total output. From the viewpoint of society, signaling is nothing but a waste of resources. However, for the members of type B, signaling is a device to get a higher share of total output. In the non-signaling-pooling case, employers cannot distinguish both worker types, *i.e.* they pay a uniform wage equal to the average productivity of all employees. By investing in education, type B avoids "subsidizing" the less productive type A. Condition (3.3) also demonstrates that there are many signaling equilibria throughout the interval $\Delta < y^* < \Delta/\mu$.[25]

---

[25] Note that the previous discussion of "uniqueness" refers to the case where the level of requirements is fixed.

The *SSE* is depicted in Fig. 3.1. The wage schedule is a bold line and jumps from a to b at $y*$. The minimum level of requirements, which might induce the *SSE*, corresponds to $y_{min} = \Delta$. The maximum level is given by $y_{max} = \Delta / \mu$. The closer the education level $y*$ is to $y_{min} = \Delta$, the lower is the welfare loss.

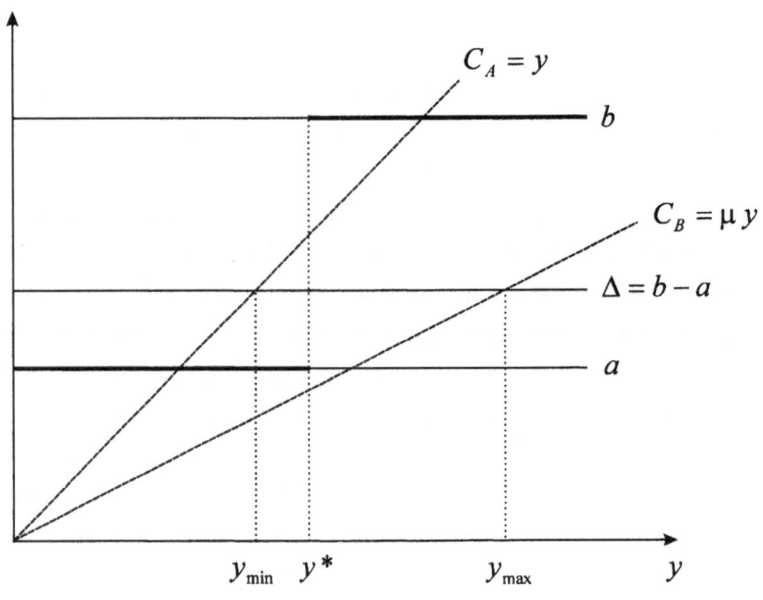

Fig. 3.1: Wage schedule and SSE

3.3   The Job-Matching Effect of Educational Signals

The productivity of an employee tends to be low if he works in an inadequate environment. Signaling may help to allocate heterogeneous individuals to their most productive use, *i.e.* it enables a correct matching of individuals to jobs. The job-matching effect and its welfare implications are formalized in this section.

3.3.1   Model Structure

Suppose that there are not only two different worker types, there are also two types of firms. The first one (*FA*) employs the workers in simple and routine jobs, and the productivity of both worker types only corresponds to *a*. But the second firm type (*FB*) offers more demanding jobs. Assume that employee type *B* can exploit its higher potential there and reaches its maximum productivity of *b*, whereas type *A* only realizes *a* again.[26] Hence, the productivity not only depends on the worker type but also on the type of employer. Again, it is assumed that the employer cannot directly observe the productivity of a single worker.

Suppose that *FB* can only hire a share $0 < p < 1$ of all job applicants. This upper bound for qualified jobs is of decisive importance for the job-matching effect. If no upper bound existed, *i.e.* $p = 1$, *FB* could hire all job applicants and the same results as in the basic model would occur. However, in the *NSPE* with $p < 1$, all employees apply for a job in *FB* because the expected payoff in *FB* exceeds the one in *FA*. Note that all type *A* employees attempt to "free-ride" on type *B*'s higher productivity in *FB*. Thus, the productive type *B* is partially driven out by type *A*. However, the signal enables the employers to distinguish the worker types, *i.e.* it prevents a misallocation of individuals to jobs.

For simplicity, assume that the proportion of qualified jobs is not less than the share of type *B* members, *i.e.* $p \geq 1 - q$. In the *SSE*, all type *B* members are identified as being productive and are hired by *FB*. Recall that the type *A* members are assumed to have the same productivity equal to *a* in both firm types. In the *SSE*, they are identified as being less productive so that they are indifferent between *FA* and *FB*.

3.3.2   Equilibrium Analysis

The self-selection condition (3.3) is still valid. In order to examine for which parameters the *SSE* is unique, assume that the level of requirements is marginally higher than the minimum value $y^* = \Delta$. If all employees decide against

---

Alternatively, we could assume that type *A*'s productivity is lower in *FB* than in *FA*. Given this assumption, the welfare-increasing effect of signaling would be even stronger. However, the computation of equilibria would be more complicated.

signaling, the probability that a job applicant is hired by *FB* is equal to *p*. The expected non-signaling payoff for an employee is given by:

$$(1-p)a + p[qa + (1-q)b].$$ (3.6)

The expected productivity in *FB* (and thus the uniform wage) corresponds to the expression in the square brackets. In order to derive conditions for a unique *SSE*, the payoff in (3.6) must be compared with type *B*'s payoff in the *SSE*:

$$b - \mu\Delta > a - pa + pqa + pb - pqb,$$ (3.7)

$$\Rightarrow \Delta - \mu\Delta > p\Delta - pq\Delta,$$

$$\Rightarrow 1 - p(1-q) - \mu > 0.$$

A single type *B* employee would always deviate from the non-signaling strategy, even if all remaining type *B* workers chose non-signaling. Since the same is true for all type *B* members, there is no *NSPE* in this situation.

In the case of $1 - p(1-q) - \mu < 0$, multiple equilibria occur. If all type *B* members select the non-signaling strategy, a single type *B* employee has no incentive to deviate because the non-signaling payoff $(1-p)a + p[qa + (1-q)b]$ exceeds the signaling payoff, which is equal to $b - \mu\Delta$. Hence, the *NSPE* exists. However, if all type *B* employees decide in favor of signaling, a single type *B* employee has no incentive to deviate from the signaling strategy because the signaling payoff $b - \mu\Delta$ is higher than the non-signaling payoff, which is then equal to *a*. Since the same is true for all type *B* employees, the *SSE* exists as well. The *SSE* can be interpreted as a coordination failure in this case because type *B* employees are better off in the *NSPE*.[27]

In the case of *NSPE*, a single worker benefits from a rising *p* because the probability increases to get a job in *FB*. The same is true for decreasing values of *q* because there are less type *A* employees to subsidize in *FB*. Surprisingly, type *B*'s productivity advantage Δ has no impact on type *B*'s decision between signaling and non-signaling, as shown in (3.7).

---

[27] In Section 3.5, we will derive the "threshold proportion" of signaling type *B* employees that makes an individual type *B* employee indifferent between signaling (*S*) and non-signaling (*NS*).

3.3.3   Welfare Implications

In order to analyze the welfare implications of job matching, we consider an additive welfare function. For simplicity, suppose that the labor market size is normalized to one. Moreover, let the requirements be on its minimum level, *i.e.* $y^* = \Delta$ so that type A is prevented from signaling.

$$W_{SSE} = q\,a + (1-q)(b - \mu\Delta). \tag{3.8}$$

In the *SSE*, both types are allocated to their most productive use because all type B employees are hired by *FB*.

In the *NSPE*, all workers apply for a job in *FB* so that type B is partially driven out by type A. Then, welfare equals the expected productivity of all employees:

$$W_{NSPE} = (1-p)a + p\big[q\,a + (1-q)b\big]. \tag{3.9}$$

Signaling is welfare-enhancing if

$$W_{SSE} > W_{NSPE}, \tag{3.10}$$

$$\Rightarrow q\,a + b - \mu\Delta - qb + q\mu\Delta > a - pa + pqa + pb - pqb,$$

$$\Rightarrow (1-q)\Delta - (1-q)\mu\Delta > (1-q)p\Delta,$$

$$\Rightarrow 1 - p - \mu > 0.$$

Note that the job-matching effect of signaling is the stronger, the lower type B's relative signaling cost μ and the lower the proportion of qualified jobs p are. In the case of $p = 1$, which corresponds to the basic model, the *SSE* is always welfare inferior to the *NSPE*. Signaling is nothing but a waste of resources because a misallocation of labor is impossible. However, with diminishing p, the probability of a misallocation increases, *i.e.* the job-matching function of signaling becomes more important. Condition (3.10) also clarifies that the parameters Δ and q do not affect the welfare comparison between the *SSE* and the *NSPE*.

Fig. 3.2 depicts the job-matching function of signaling for the numerical example $p = 0.5$, $q = 0.8$, $\mu = 0.5$ $a = 1$ and $b = 2$. In the *SSE*, all type B workers are hired by *FB* where they can reach their maximum productivity of b.

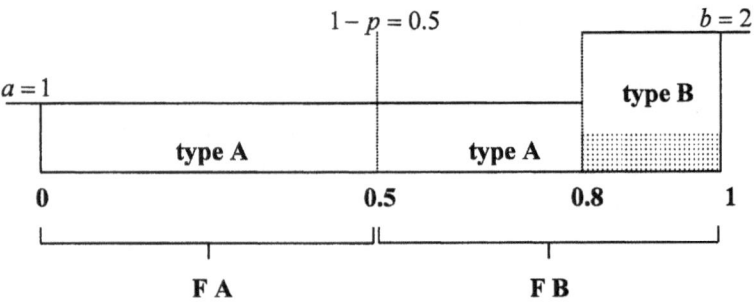

Fig. 3.2: The SSE

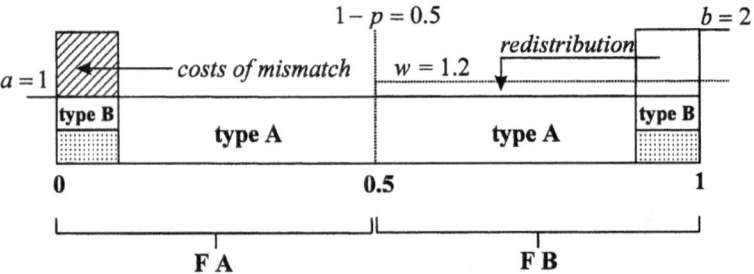

Fig. 3.3: The NSPE

Using the same numerical example, Fig. 3.3 illustrates the *NSPE*. It is shown that half of the type *B* employees are driven out by type *A*. The hatched area in Fig. 3.3 stands for the expected loss that would be caused by mismatch. It can be prevented by signaling. In the borderline case $1 - p = \mu = 0.5$, this welfare loss is equal to the signaling costs represented by the spotted area. The arrow within *FB* indicates the income that is redistributed within *FB*: Since all employees within *FB* get uniform wages of $\overline{w} = 1.2$, type *B* "subsidizes" type *A*.

In Fig. 3.4, it is assumed that $y*$ is always on its minimum level ($y* = \Delta$) and that $q = 0.5$ holds. The spotted area represents the region of the unique *SSE* that is welfare inferior. This inefficiency can be traced back to the fact that the type *B* employees also internalize the welfare neutral distributive effect of signaling. For them, signaling is a device to avoid subsidizing type *A*. But from the welfare perspective, signaling is not desirable in this area be-

cause signaling costs exceed the positive job-matching effect. This inefficiency area is given by

$$2\mu - 1 < 1 - p < \mu.\qquad(3.11)$$

To the right of this area, there are multiple equilibria, *i.e.* both the *SSE* and the
NSPE are feasible, and the *SSE* can be interpreted as a coordination failure
because the *NSPE* would make each type *B* (and type *A*) employee better off.

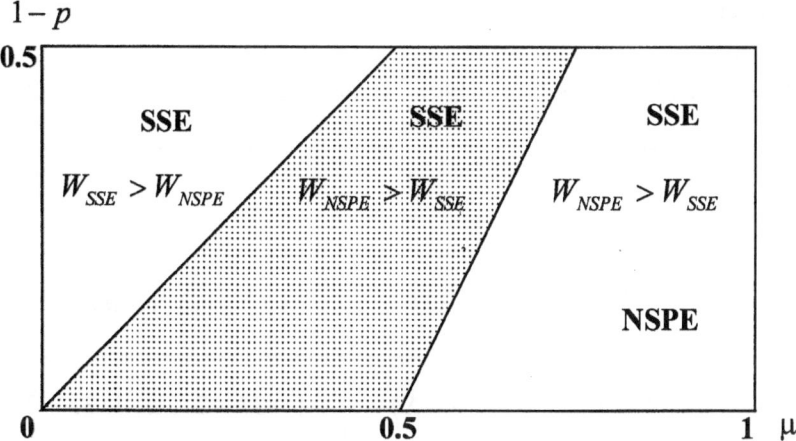

Fig. 3.4: Equilibria and welfare areas

3.3.4   Job Matching with a Small Number of Employees

Assume that the employers are able to measure the total ex-post output but
that they are unable to observe the contribution of a single employee. If there
are only few employees, each has a significant impact on total output. This
impact may be an alternative way to "signal" the own type.

Suppose that there are only two type *A* and two type *B* employees. Let the
level of requirements be marginally higher than $y^* = \Delta$ so that both type *A*
employees are prevented from signaling. Moreover, assume that there are only
two jobs in *FB*, *i.e.* $p = 1/2$. Matrix 3.1 depicts the payoffs for both type *B*
employees. They can choose between the strategies "signaling" (*S*) and "non-

signaling" (NS). The employers are assumed to pay a wage equal to the expected productivity of the employee.

The payoffs for the strategy combinations (NS, S) and (S, NS) are computed as follows: The signaling type B employee can be identified as being productive and thus obtains a wage equal to b. The probability that the non-signaling type B employee is hired by FB is 1/3. In this case, the total output of FB equals 2b and even though the employee has chosen NS, he receives a wage of b because the employer identifies him as type B. But with probability 2/3 he is merely hired by FA, where he gets a wage of a. Hence, the expected payoff for the non-signaling type B employee is equal to $1/3b+2/3a$.

Matrix 3.1: Payoffs of the two type B employees

	S_2	NS_2
S_1	$(b-\mu\Delta;\ b-\mu\Delta)$	$(b-\mu\Delta;1/3b+2/3a)$
NS_1	$(1/3b+2/3a;\ b-\mu\Delta)$	$(1/3b+2/3a;\ 1/3b+2/3a)$

In the case of (NS, NS), all job applicants attempt to get a job in FB. The expected payoff for a type B employee corresponds to

$$1/2[1/3b+2/3(a+b)/2]+1/2a=1/3b+2/3a.$$

With probability 1/2 the employee is hired by FB. There is a chance of 1/3 that he meets the other type B employee in FB and gets a wage equal to b. However, his probability is 2/3 to become the colleague of a type A employee. In this case, he would only get $(a+b)/2$. In the worst case, the employee is hired by FA, where his wage will be a. The probability of this event is 1/2.

The game has the following Nash-equilibria: For $\mu > 2/3$, (NS, NS)* is a unique equilibrium. If $\mu < 2/3$, (S, S)* is a unique equilibrium.

Unlike the model with a larger number of employees, (NS, NS) can be a unique equilibrium. Given the assumption of a large number of employees, it is never beneficial for a single type B employee to choose NS if the remaining type B members choose S because the expected productivity of a non-signaling employee equals a in this situation. However, if we assume a small number of employees, each employee has a significant impact on the non-signaling payoff. For example, if both type B employees meet in FB, they get b even if they have chosen NS before. Thus, the type B employees are able to "signal" their higher productivity in FB by their impact on total output. With increasing signaling costs μ, this alternative way of "signaling" becomes more attractive.

In the SSE, both type B employees are hired by FB, whereas the two type A employees only get a job in FA. Welfare corresponds to

$$W_{SSE} = 2(b - \mu \Delta) + 2a.$$

In the NSPE, welfare just corresponds to the expected total output

$$W_{NSPE} = 1/6 \cdot 2b + 2/3\,(a+b) + 1/6 \cdot 2a + 2\,a = 3\,a + b.$$

With probability 1/6 both type B employees meet in FB, where they have a total output of $2b$. The probability equals 2/3 that a type B employee meets a type A employee in FB. In this case, the total output of FB corresponds to $a+b$. With probability 1/6 both type A employees are hired by FB, where they have a total output of $2a$. The two employees in FA produce $2a$ altogether irrespective of their type. Signaling is welfare-increasing if the following condition holds:

$$W_{SSE} > W_{NSPE} \qquad \Rightarrow \mu < 1/2.$$

Fig. 3.5 depicts the equilibrium and welfare areas for this example. The spotted area represents the parameter combinations for a unique signaling equilibrium that is welfare inferior. From the theoretical welfare perspective, signaling is not desirable in this area because the costs of signaling exceed the welfare-enhancing job-matching effect.

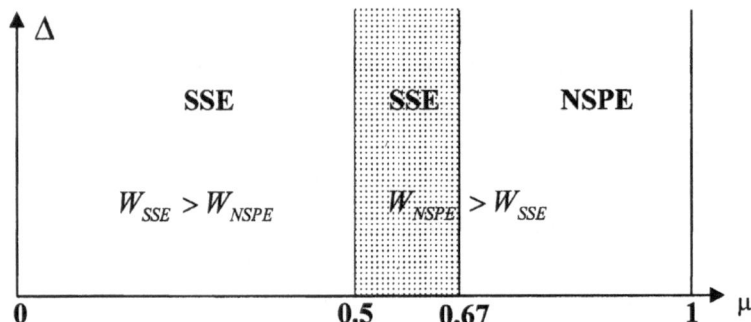

Fig. 3.5: Equilibrium and welfare areas

3.4   Standardization of Examination Requirements

If examination requirements vary among universities, it might be beneficial for a type A employee who faces low requirements to acquire the educational signal, *i.e.* the diploma. On the other hand, a type B employee who is confronted with relatively high requirements could be willing to leave the university without graduating. Thus, there would not exist a "pure" *SSE* any longer and the job-matching effect of signaling would deteriorate. By standardizing the requirements, for instance by means of central examinations or accreditation, signaling could recover its filtering function.

3.4.1   Model Structure

Suppose that for both worker types, the requirements y are uniformly distributed in the interval $[m-d, m+d]$, where m stands for the mean value and d denotes the deviation with $0 \leq d \leq m$. For $d = 0$, there is full standardization of requirements because the educational signal can only be acquired with a single level of $y = m$. Then, the signal is perfect and the employers identify signaling employees as type B. But with increasing d, the signal's quality is reduced.

The model has two stages. Suppose that in the first stage, all employees take up their studies. This decision is based upon expected values. Note that in the first stage, employees of the same type are homogeneous. In the second stage, each employee finds out his actual level y_i. Requirements y are modeled as a kind of experience good, *i.e.* after a certain time, requirements can be assessed by the students. Interpreting the allocation of y as a random selection, other possible determinants of the allocation are ruled out. For example, students may have different private information about requirements and/or a different degree of mobility. Moreover, students who face relatively high levels of requirements could switch to another university.

An employee who faces relatively low requirements decides in favor of signaling at costs y_i or μy_i, respectively. If the requirements are relatively high, employee i leaves the university at costs $\tilde{y}_i < y_i$ without graduating, *i.e.* without signal. For simplicity, suppose that the costs for non-graduating are equal to zero, *i.e.* $\tilde{y}_i = \tilde{y} = 0$. A share of type A employees could be able to acquire the signal, whereas a proportion of type B might leave university without graduating because of high signaling costs. It is assumed that the employers are not able to distinguish different levels of requirements. Thus, universities do not build up reputation. However, the distribution of y is assumed to be common knowledge so that everybody can derive the probability that a signaling employee belongs to type B or type A, respectively. Let the employers pay a wage that corresponds to the employee's expected productivity.

3.4.2 Bayesian Equilibria

Fig. 3.6 depicts the four possible subsets of a Bayesian equilibrium. For instance, the subset $B \cap S$ represents the type B employees who acquire the signal. The equilibrium proportion of signaling type B employees to all type B employees corresponds to $P(S|B)^*$. Analogously, the subset $A \cap S$ stands for the signaling type A employees, and $P(S|A)^*$ denotes the equilibrium proportion of signaling type A employees to all type A employees.

It is straightforward to derive the equilibrium proportion of signaling employees to all employees:

$$P(S)^* = P(A \cap S)^* + P(B \cap S)^* = q\, P(S|A)^* + (1-q)\, P(S|B)^*. \qquad (3.12)$$

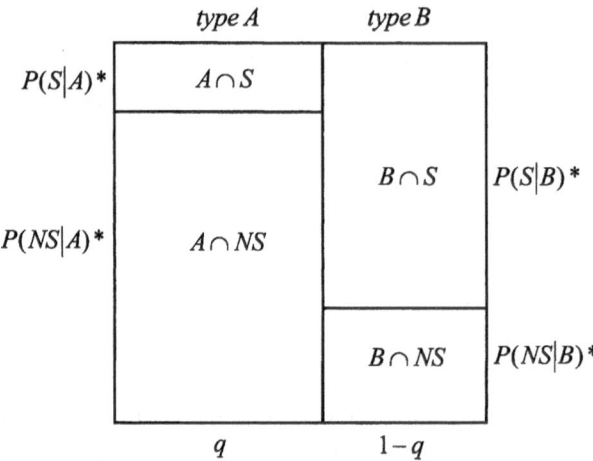

Fig. 3.6: Subsets of a Bayesian equilibrium

Assume that the total number of employees is equal to the total number of jobs. To keep things simple, suppose again that the labor market size is normalized to one.

Applying the Bayesian rule yields $\lambda^* = P(B|S)^*$ as the equilibrium probability that a signaling employee belongs to type B.

$$\lambda^* = P(B|S)^* = \frac{P(B \cap S)^*}{P(A \cap S)^* + P(B \cap S)^*} = \frac{(1-q)\,P(S|B)^*}{P(S)^*}. \qquad (3.13)$$

Note that λ^* can be interpreted as a measure of the signal's quality because in the case of full standardization ($d = 0$), λ^* is equal to one.

The subsets $A \cap NS$ and $B \cap NS$ denote the non-signaling type A and type B employees. The equilibrium probability that an employee without signal belongs to type B, i.e. $\psi^* = P(B|NS)^*$, is given by:

$$\psi^* = P(B|NS)^* = \frac{P(B \cap NS)^*}{P(A \cap NS)^* + P(B \cap NS)^*} = \frac{(1-q)(1-P(S|B)^*)}{1-P(S)^*}. \qquad (3.14)$$

In the following, let us distinguish two cases. In the first one, the proportion of jobs in *FB*, p, exceeds the equilibrium proportion of signaling employees $P(S)^*$. Consequently, *FB* hires all signaling employees and allocates the remaining places to non-signaling employees. In the second case, there are more signaling employees than jobs in *FB*.

Case 1 ($P(S)^* < p$): Player i of type A chooses S if

$$\lambda^* b + (1-\lambda^*)a - y_{A,i} > \frac{p-P(S)^*}{1-P(S)^*}\big(\psi^* b + (1-\psi^*)a\big) + \frac{1-p}{1-P(S)^*}a. \qquad (3.15)$$

The left-hand side represents the employee's signaling payoff. The signaling employee obtains a wage equal to his expected productivity $\lambda^* b + (1-\lambda^*)a$. The right-hand side of (3.15) denotes the expected non-signaling payoff. All non-signaling employees apply for a job in FB, because in this firm type, they get a wage equal to their expected productivity $\psi^* b + (1-\psi^*)a$, which exceeds the uniform wage a, paid by FA. With probability $(p-P(S)^*)/(1-P(S)^*)$, a non-signaling employee is hired by FB. With the inverse probability $(1-p)/(1-P(S)^*)$, the employee is hired by FA.

Solving (3.15) for y_A yields:

$$y_A^* = \frac{\psi^*(1-p)\Delta}{1-P(S)^*} + \Delta(\lambda^* - \psi^*). \qquad (3.16)$$

The type A employee who is located at y_A^* is indifferent between S and NS. For $y_{A,i} < y_A^*$, a type A employee chooses S. But in the case of $y_{A,i} > y_A^*$, it is not beneficial for a type A member to acquire the signal.

For type B, the indifference level of requirements is computed analogously:

$$y_B^* = y_A^*/\mu. \qquad (3.17)$$

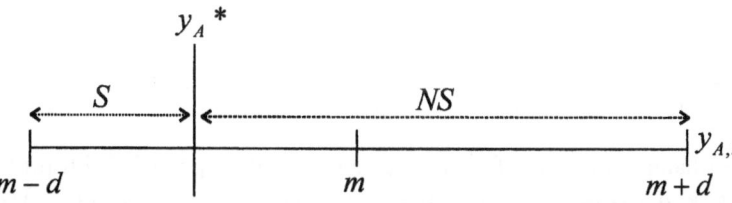

Fig. 3.7: Proportion of signaling type A employees to all type A workers

Fig. 3.7 depicts the equilibrium proportion of signaling type A employees to all type A employees. If the indifference level y_A^* is lower than the minimum value of requirements, $m-d$, no type A employee will acquire the signal. However, all type A employees choose signaling if y_A^* exceeds the maximum level of requirements $m+d$. If the indifference level is within the interval of possible requirements, as shown in Fig. 3.7, the distance between the indifference value and the minimum level, $m - d$, has to be divided by the length of the whole interval, which equals $2d$.

The equilibrium proportion of signaling type A employees to all type A employees corresponds to:

$$P(S|A)* = \begin{cases} 0, & \text{if} \quad y_A* < m - d \\ \dfrac{y_A* - m + d}{2d}, & \text{if} \quad m - d \leq y_A* \leq m + d \cdot \\ 1, & \text{if} \quad m + d < y_A* \end{cases} \tag{3.18}$$

Analogously, the equilibrium proportion of signaling type B employees to all type B members is given by:

$$P(S|B)* = \begin{cases} 0, & \text{if} \quad y_A*/\mu < m - d \\ \dfrac{y_A*/\mu - m + d}{2d}, & \text{if} \quad m - d \leq y_A*/\mu \leq m + d \cdot \\ 1, & \text{if} \quad m + d < y_A*/\mu \end{cases} \tag{3.19}$$

Proportions (3.18) and (3.19) describe the equilibrium outcome. In Section 3.4.4, we will compute these values for a numerical example.

Case 2 ($p < P(S)*$): Here, the proportion of signaling employees exceeds the proportion of jobs in FB. Even if an employee acquires the signal, he may not get a job in FB. An employee of type A chooses S if

$$\frac{p}{P(S)*}(\lambda* b + (1 - \lambda*)a) + \frac{P(S)* - p}{P(S)*}a - y_{A,i} > a . \tag{3.20}$$

The left-hand side stands for the expected signaling-payoff. With probability $p/P(S)*$ the employee gets a job in FB. The fraction $(P(S)* - p)/P(S)*$ denotes the probability to be hired by FA. Solving condition (3.20) for $y_{A,i}$ yields type A's indifference value of requirements:

$$y_A* = \frac{\lambda* p\Delta}{P(S)*} . \tag{3.21}$$

The indifference level for type B is given by

$$y_B^* = y_A^* / \mu = \frac{\lambda^* p \Delta}{\mu P(S)^*}.$$

(3.22)

Conditions (3.12), (3.13), (3.18), and (3.19) are valid for case 2 as well.

3.4.3   Welfare Implications of Standardization

In order to analyze the welfare implications of standardization, we distinguish again the two cases. Welfare is derived by means of an additive welfare function, which sums up the expected total output and total signaling costs.

Case 1 ($P(S)^* < p$): The expected total output X^* is given by

$$X^* = P(S)^* \left[\lambda^* b + (1 - \lambda^*) a \right] \\ + \left(p - P(S)^* \right) \left[\psi^* b + (1 - \psi^*) a \right] + (1 - p) a .$$

(3.23)

All signaling employees are hired by *FB*. The remaining places in *FB* are allocated to non-signaling employees.

Parameter C_A^* stands for type *A*'s total costs of signaling.

$$C_A^* = \begin{cases} 0, & \text{if } y_A^* < m - d \\ q P(S|A)^* \dfrac{y_A^* + m - d}{2}, & \text{if } m - d \leq y_A^* \leq m + d . \\ q m, & \text{if } m + d < y_A^* \end{cases}$$

(3.24)

If y_A^* is within the interval of possible requirements, $m - d < y^* < m + d$, the proportion of signaling type *A* employees is equal to $P(A \cap S) = q P(S|A)^*$. For computation of C_A^*, this proportion must be multiplied with the average level of requirements for signaling type *A* employees, which is equal to $(y^* + m - d)/2$. If y_A^* exceeds the maximum level of requirements, $m + d$, all type *A* employees acquire the signal and total signaling costs correspond to $q m$.

The expected signaling costs for type *B* are computed analogously:

$$C_B^* = \begin{cases} 0, & \text{if } y_A^* / \mu < m - d \\ (1 - q) P(S|B)^* \mu \dfrac{y_A^* / \mu + m - d}{2}, & \text{if } m - d \leq y_A^* / \mu \leq m + d . \\ (1 - q) \mu m, & \text{if } m + d < y_A^* / \mu \end{cases}$$

(3.25)

Then, welfare is given by:

$$W* = X* - C_A* - C_B*.$$ (3.26)

Case 2 ($p < P(S)*$): In this case, the expected total output corresponds to

$$X* = p[\lambda*b + (1-\lambda*)a] + (1-p)a.$$ (3.27)

All jobs in *FB* are allocated to signaling employees. Their expected productivity is given by $\lambda*b + (1-\lambda*)a$. Conditions (3.24), (3.25), and (3.26) are still valid.

3.4.4   Numerical Example

The equilibrium probabilities of the above model can only be derived numerically. Suppose that $q = p = 0.5$, $m = 1$, $\mu = 0.6$, $b = 2$, and $a = 1$ are given.

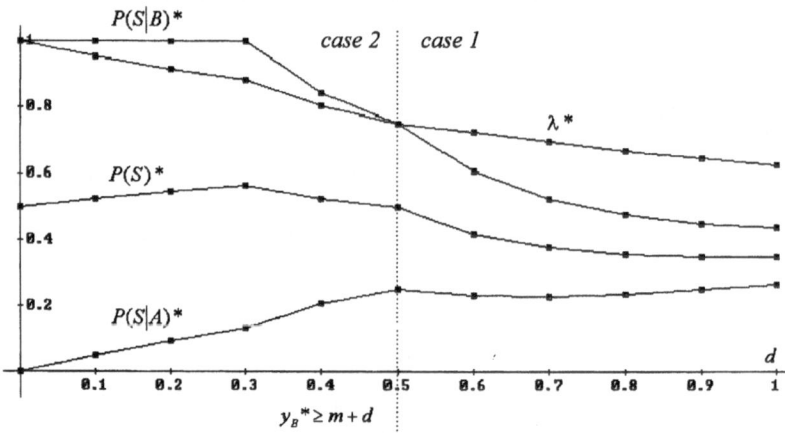

Fig. 3.8: Bayesian equilibrium depending on the deviation of requirements

Fig. 3.8 depicts the equilibrium values of $P(S|A)*$, $P(S|B)*$, $\lambda*$, and $P(S)*$ depending on the degree of deviation. The equilibrium values are computed numerically for $d = \{0, 0.1, 0.2, 0.3, 0.4, 0.5, 0.6, 0.7, 0.8, 0.9, 1\}$. The approximated solutions for case 1 and case 2 are listed in appendix A 1.

As shown in Fig. 3.8, case 1 holds for $0.5 < d \leq 1$ and case 2 for $0 \leq d < 0.5$. Whereas $y_A{}^*$ always remains within the interval of possible requirements, $m - d \leq y_A{}^* < m + d$, type B's indifference value, $y_B{}^* = y_A{}^* / \mu$, exceeds maximum requirements for $d \leq 0.299603$. Consequently, for low deviations, all type B employees acquire the signal, i.e. $P(S|B)^* = 1$. It is interesting to note that $P(S|A)^*$ and $P(S)^*$ have local maxima for relatively low values of d, whereas $P(S|B)^*$ and λ^* never increase with d. The uniqueness of the Bayesian equilibrium can be traced back to the fact that condition (3.7) holds in this numerical example, i.e. in the case of full standardization, the *SSE* is unique.

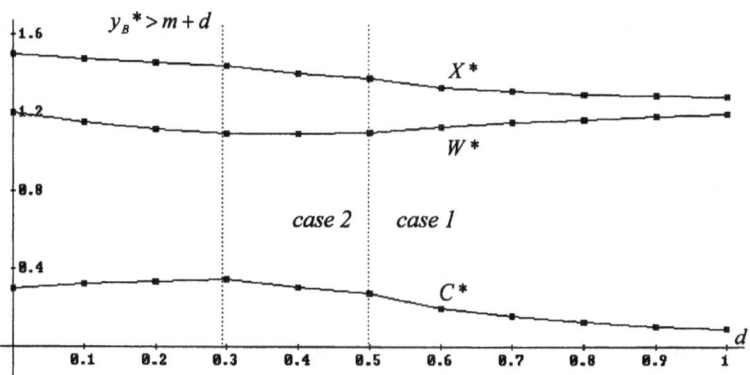

Fig. 3.9: Welfare implications of standardization

Fig. 3.9 depicts X^*, C^*, and W^* depending on the degree of deviation d. The approximated solutions are listed in appendix A 1. If d increases, the expected total output X^* declines due to the deterioration of job matching. The impact on total signaling costs is ambiguous: For low deviations, $0 < d < 0.299603$, total signaling costs increase in d, but for levels of deviations within interval $0.299603 \leq d \leq 1$, they decrease. Hence, there is a trade-off between X^* and C^* in this case. The welfare function is u-shaped, i.e. the case of full standardization ($d = 0$) has to be compared with the case of full deviation ($d = 1$). In our example, full standardization is welfare superior to full deviation, i.e. $\mathrm{W}(d = 0)^* = 1.2 > \mathrm{W}(d = 1)^* = 1.191718$. The intermediary cases $0 < d < 1$ turn out to be welfare inferior.

For given values of $a = 1$, $b = 2$, $m = 1$, and $q = p = 0.5$, full standardization is equivalent to full deviation for $\hat{\mu} = 0.611883$. If μ exceeds this threshold value, total signaling costs outweigh the positive job-matching effect. Then, society is better off with full deviation.

3.5   Equilibrium in Mixed Strategies for the Case of Perfect Signals

In Section 3.3, we have shown that multiple equilibria exist if $d = 0$. Then, the decision of a single type B employee depends on the strategy selected by the other type B members. If all type B employees choose NS, it is not beneficial for a single type B employee to deviate. But if the type B employees unanimously decide in favor of S, the best response for a single type B member is S as well. Suppose there is a proportion of type B, $0 \le P(S|B) \le 1$, selecting strategy S. What would be the best response for a single type B employee? Which proportion $P(S|B)^*$ would make a single type B member indifferent between the strategies S and NS? In order to answer these questions, we will hark back to the approach from Section 3.4.

Let the requirements be on its minimum level $y^* = \Delta$ so that no type A employee chooses S, i.e. $P(S|A) = 0$. Note that case 1 applies due to $p \ge 1 - q > P(S) = (1 - q) P(S|B)$. Making use of the right-hand side of (3.15) yields the non-signaling payoff for a single type B employee:

$$EU_{NS} = \frac{p - P(S)}{1 - P(S)} \left[\psi b + (1 - \psi) a \right] + \frac{1 - p}{1 - P(S)} a. \tag{3.28}$$

Making use of (3.12) and (3.14) and substituting $P(S|A) = 0$ yields:

$$EU_{NS} = b + \frac{(1 - p) q \Delta}{\left[1 - (1 - q) P(S|B) \right]^2} - \frac{(1 - p + q) \Delta}{1 - (1 - q) P(S|B)}. \tag{3.29}$$

Fig. 3.10 shows that the expected payoff for the non-signaling strategy decreases in $P(S|B)$. For $P(S|B) = 0$, EU_{NS} corresponds to $(1 - p) a + p \left[q a + (1 - q) b \right]$. In the case of $P(S|B) = 1$, the non-signaling employee obtains a. EU_{NS} intersects $b - \mu_2$ in $P(S|B)^* = 0.5375$. If a share of type B workers less than $P(S|B)^*$ chooses S so that NS is the best response for a single type B employee, the game converges to the equilibrium in which everyone chooses *NS*. For more than $P(S|B)*$ percent playing S, the best response for a single type B employee is S so that the game converges to the *SSE*.

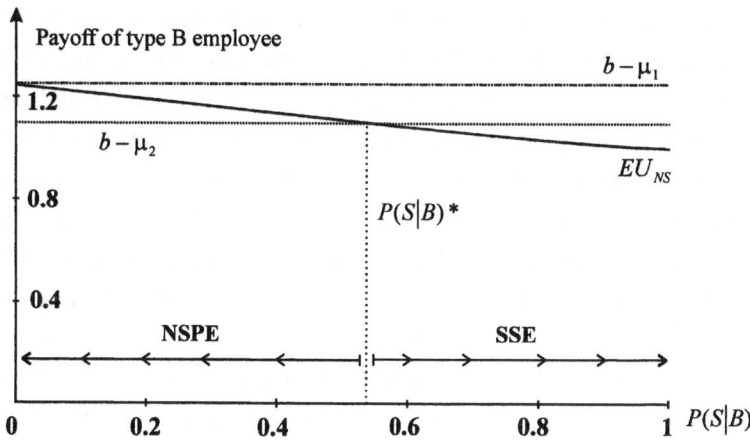

Fig. 3.10: Equilibrium in mixed strategies for parameters $q = p = 0.5$, $b = 2$, $a = 1$, $\mu_1 = 0.75$, $\mu_2 = 0.9$.

An equilibrium in mixed strategies implies that all type B employees randomize with $P(S|B)*$ between strategies S and *NS*. Since the expected proportion of signaling type B employees is then equal to $P(S|B)*$, the single type B member is indifferent between S and *NS*.

If signaling costs decrease to μ_1, the signaling-payoff function shifts up. In Fig. 3.10, both strategies have the same (expected) payoffs at $P(S|B)* = 0$. Thus, in the case of $0 < P(S|B) \leq 1$, a single type B employee always chooses signaling. Note that μ_1 defines the borderline case $1 - p(1-q) - \mu_1 = 0$. It follows from (3.7) that the *SSE* is unique for $\mu < \mu_1$, *i.e.* an equilibrium in mixed strategies is not feasible any longer.

3.6 Conclusions

The job-matching approach has shown that signaling may increase total output by allocating heterogeneous employees to the adequate firm type. Hence, we have derived the result that education can be welfare-enhancing without assuming that education raises the students' human capital. If the separation between the signaling and human capital theory is removed[28], private and social benefits of education increase even more.

The job-matching approach has demonstrated that private incentives for signaling are too strong. This inefficiency may occur because the type B employees also internalize the welfare-neutral distributive effect of signaling. For them, signaling is a device to get a higher share of the total output because they avoid subsidizing the less productive type. The equilibrium analysis has shown that, given the self-selection conditions hold, the *SSE* is unique for relatively low signaling costs. If signaling costs are relatively high, multiple equilibria exist, *i.e.* the *SSE*, the *NSPE*, and an equilibrium in mixed strategies are feasible. Since the *NSPE* would make all type B employees better off in this situation, the *SSE* can be interpreted as a coordination failure. For a small number of employees (two type A and two type B employees) a unique *NSPE* does exist because the single type B employee can "signal" his higher productivity by means of his influence on total output.

In the second part of this chapter, we have introduced incomplete information about the signal's quality. We have made the assumption that a continuum of examination requirements exists. Standardization implies that the range of requirements is reduced, for instance by means of central examinations or accreditation. As shown in the numerical analysis, a unique Bayesian equilibrium exists for different levels of deviations d. The welfare analysis has revealed that standardization of the requirements increases the expected total output by improving the job-matching function of signaling. However, standardized requirements may also increase expected total signaling costs. Welfare effects of standardization are ambiguous: For low signaling costs, full standardization is desirable. If the signaling costs are relatively high, society is better of with a full range of requirements, *i.e.* with maximal deviation.

---

[28] See, for example, Riley (1976) who combines the signaling approach with the human capital theory. The human capital approach goes back to Becker (1964).

4

Inter-Technology versus Intra-Technology Competition in Network Markets

This chapter analyzes the competition between two firms when their incompatible technologies exhibit network effects in that high expected sales increase the willingness to pay for the corresponding good. An incumbent firm faces the strategic choice of whether to share its superior technology (via free licensing) with a follower or to keep its technology for itself. The first option of sponsoring intra-technology competition increases the incumbent firm's network and thus consumers' willingness to pay. On the other hand, the latter option involves inter-technology competition. Depending on the relative cost advantage of the incumbent firm, the entry of the rival technology may be blockaded, both technologies can coexist in an incompatible duopoly, or the incumbent firm may deter the entry of its rival. The model investigates the incumbent firm's choice of whether to sponsor intra-technology competition or to insist on inter-technology competition.[29]

4.1 Introduction

A fundamental question of strategy for a firm facing horizontal competition in a network market is whether to share its technology with other firms and to compete within a joint network or to keep its technology for itself.[30] With just two firms, this strategic choice involves three combinations of strategies (Besen and Farrell, 1994). In the first, both firms prefer inter-technology compe-

---

[29] This chapter builds upon Langenberg (2005).
[30] Alternatively, firms could ensure (at least partial) compatibility by using converters. In this case, they might adopt different technologies without losing the network advantage of a joint technology. For this alternative, see Farrell and Saloner (1992).

tition to determine the industry standard. In the second case, which corresponds to the Battle of the Sexes, both firms want to compete within a joint network, but they cannot agree on the standard: "Each sponsor wants the other to join its network but would be willing to join the other's if the alternative is incompatibility" (Besen and Farrell, 1994, p. 125). Finally, in the last case, one sponsor may prefer to maintain its technology while the competitor may wish to join the rival's network.

It is an empirical question whether firms actually invite potential rivals to license their technology and to compete within a joint network. In the early 1980s, Intel licensed its microprocessor designs to AMD. Later on, Intel reversed this decision and broke off its agreement with AMD (see, for example, Shapiro and Varian, 1999, p. 125). Another well-known example for both inter-technology and intra-technology competition is the VHS/Betamax contest to determine the standard for video cassette recorders. Despite an early lead and a superior quality, the Sony Betamax system was finally driven out of the market by the VHS system from Japan Victor Corporation (JVC). The ultimate victory of VHS can be attributed to JVC's strategy of licensing its technology to other manufacturers (see, for example, Grindley, 1995, pp. 75-130).

In this model, we follow Economides (1996a) who assumes that high expected sales increase the willingness to pay for the network good. He shows that it can be beneficial for the exclusive holder of a network technology to invite competitors into its network. The reason for this "seeming paradox" (Economides, 1996a, p. 31) is that the holder of the technology cannot credibly commit to a network size which exceeds its relatively low profit-maximizing monopoly output. Consumers' expectations have to be fulfilled at the equilibrium level. The invitation of other firms is a self-binding device for the innovator because rational consumers anticipate that the market's total output rises with the number of competitors.[31]

Whereas the model by Economides is confined to intra-technology competition, this model also deals with inter-technology competition. We focus on two firms A and B, each is the exclusive holder of a technology. The modeling of inter-technology competition is based on a system of two linear demand

---

[31] See Farrell and Gallini (1988) for an earlier model analyzing monopoly incentives to attract competition.

functions. Total demand can be increased by network effects, *i.e.* the market size is not fixed such as in models of the Hotelling-type. Two asymmetries occur: First, firm A is assumed to be the quantity leader. Secondly, the technologies may have different marginal costs. The incumbent firm A faces the problem of whether to share its superior technology (via free licensing) with the potential competitor B or to keep its technology for itself. Like in the model by Economides, the invitation of competitor B into product market A increases the network size and thus consumers' willingness to pay.

If the incumbent keeps its technology for itself, three different equilibria may occur. Depending on the relative cost advantage of technology A, the entry of technology B can be blockaded[32], both technologies may coexist in an incompatible and heterogeneous duopoly or the incumbent can deter the entry of technology B. We will show that the incumbent can realize a higher profit in the case of entry deterrence than in the situation where the competitor's entry is blockaded.[33] Conversely, the textbook result (see, for example, Pfähler and Wiese, 1998, p. 116) is that the incumbent's entry deterrence strategy is less profitable. By supplying the deterrence output, the incumbent not only prevents the competitor's entry, it also credibly commits to a quantity which exceeds the profit-maximizing monopoly output. As a consequence, consumers have an increased willingness to pay. Furthermore, it is shown that the deterrence profit can rise with decreasing marginal costs of the competing technology. The reason for this perverse effect is that the positive network effect of expanding the quantity may exceed the deterrence costs.

There already exist models dealing with network effects and entry deterrence. Generally, these models examine how an incumbent exploits network effects in order to prevent the entry of a competing technology. For example, Church and Gandal (1996) assume that the incumbent can commit earlier to an installed base of complementary goods than its competitor. By over-investing

---

[32] In Section 4.3, we will omit the assumption that technology A is superior. As a consequence, the entry of A can be blockaded if technology B has a considerable cost advantage.

[33] In this essay, the usual terminology is used which distinguishes between "deterred" and "blockaded" entry. Blockaded entry occurs when the incumbent just chooses the profit-maximizing strategy and, given this decision, it is not worthwhile for the competitor to enter the market. Entry deterrence refers to situations where the incumbent has to modify its strategy in order to prevent entry because the competitor is not as weak in terms of costs or quality as in the former case.

in the installed base, the incumbent strengthens the indirect network effects for its technology so that the market entry of the competing system is prevented. In an earlier model, Farrell and Saloner (1986a) demonstrate that the incumbent can prevent the entry of a competing network technology by using predatory pricing.[34]

In this model, it will be demonstrated that entry deterrence can only be a fulfilled expectations equilibrium if the incumbent's cost advantage is not too strong. In the case of a strong cost advantage, the entry of the competing technology is blockaded and the incumbent cannot credibly commit to a network size that exceeds its profit-maximizing monopoly quantity. On the other hand, if one technology only has a weak cost advantage, both technologies coexist in an incompatible and heterogeneous duopoly.

The requirement, that expectations have to be fulfilled in the equilibrium, makes it necessary to determine the equilibrium in a two-step approach: In the first step, we will compute the profit-maximizing quantities that fulfill expectations for a given path. The second step deals with the fact that firms could leave the path. Stability conditions will be derived for each path and it will be shown that these conditions ensure the uniqueness of the fulfilled expectations equilibrium.[35]

The structure of the model is as follows: Section 4.2 presents the model assumptions. Section 4.3 is confined to inter-technology competition. Section 4.4 extends the analysis to the incumbent's strategic problem of whether to share its technology with the competitor or to keep its technology for itself. In order to focus on the comparison of entry deterrence with the invitation strategy, we will assume that technology A has a moderate cost advantage. It will be demonstrated that entry deterrence can be more profitable for the incumbent than inviting entry. Concluding remarks follow in Section 4.5.

---

[34] See Matutes and Regibeau (1996) for a review.

[35] Here, the term "uniqueness" refers to the case of inter-technology competition. The incumbent's strategic decision between inter-technology and intra-technology competition involves multiple equilibria.

4.2 Model Structure

Suppose that there are two incompatible network technologies $i = A, B$ exclusively held by two firms. The demand for the network goods A and B is given by the following system of linear inverse demand functions:

$$p_i = \alpha_i - \beta_i y_i - \gamma y_j + n y_i^e, \qquad i, j = A, B, \qquad i \neq j. \tag{4.1}$$

The parameter γ denotes the degree of substitutability.[36] For simplicity, suppose that $\beta_A = \beta_B = 1$ holds. Moreover, let us assume $0 < \gamma < 1$, thus the markets for A and B are interdependent. Network effects for good i depend on the level of expected sales y_i^e and on the network effect parameter $0 \leq n \leq 1$.[37] Thus, network effects push the demand curves (for given expectations) up without changing their slopes.[38] For each unit sold, the network benefit is the same. For simplicity, suppose that technologies have constant marginal costs with $0 \leq c_i < \alpha_i$. Hence, at least one firm will supply a positive quantity.

The model has three stages. In the first stage, consumers form expectations by backward induction, *i.e.* they anticipate the equilibrium levels of sales. In the second stage, firm A (which is assumed to be the quantity leader) selects the profit-maximizing quantity of A. Firm B responds in the third stage with the profit-maximizing quantity of B. Four equilibrium paths may occur:

1. Both technologies can coexist in a heterogeneous and incompatible duopoly, *i.e.* firm B responds with $y_{B,I}^e{}^* > 0$ to the leader quantity $y_{A,I}^e{}^* > 0$.

2. The market entry of firm B can be blockaded if firm A has a strong cost advantage. Then, firm A is able to select the profit-maximizing quantity $y_{A,M}^e{}^* > 0$ without paying attention to firm B.

3. In the case of a strong cost advantage of technology B, the market entry of technology A may be blockaded so that follower B selects the profit-maximizing quantity $y_{B,M}^e{}^* > 0$ in the third stage.

---

[36] This linear demand system – except the network effect – goes back to Spence (1976), Dixit (1979), and Singh and Vives (1984).

[37] For a discussion of the functional form of network effects , see Swann (2002).

[38] See Wiese (1997) for the distinction between the "demand curves for a given level of sales expectations" and the "fulfilled expectations demand curve", which is defined by equating expected sales with actual demand.

4. If firm A has a moderate cost advantage, it can exploit its first mover advantage in order to deter the entry of the competing technology.[39] The deterrence quantity $y_{A,D}^{e}* > 0$ is derived from firm B's reaction function. It corresponds to the quantity of good A which makes firm B respond with an output equal to zero.

Fig. 4.1 summarizes the timing structure and the possible equilibrium paths which must fulfill consumers' expectations.

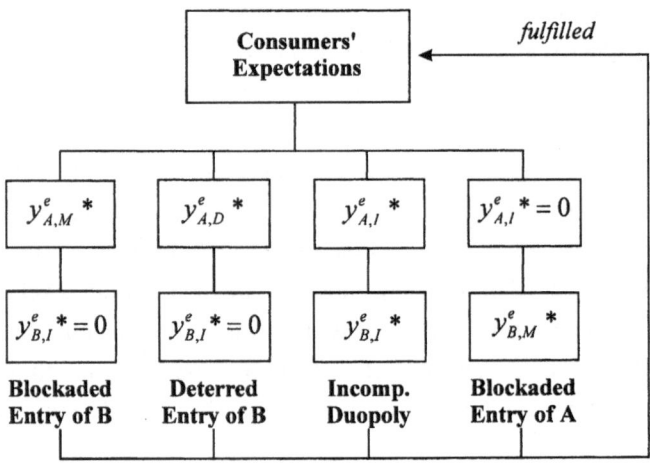

Fig. 4.1: Model structure and possible equilibrium paths

4.3   Fulfilled Expectations Equilibrium

The fulfilled expectations equilibrium is computed in a two-step approach. In the first step, the profit-maximizing quantities are derived for a given path, *i.e.* the quantities must be equal to the path expectations. For example, consider the case where firm A has a monopoly and the entry of technology B is block-

---

[39]   We do not distinguish between „quantity" and „capacity" in this model. However, Dixit (1980) demonstrates in a two-stage game that the incumbent cannot deter entry by investing in a large capacity, *i.e.* the deterrence strategy would not be subgame-perfect.

aded. As usual, we apply the backward-induction principle. Starting in the second stage, the profit-maximizing monopoly output is computed as a function of the expected network size. Finally, in the first stage, the expectations are endogenized: At the equilibrium level, actual quantities have to be equal to the expected ones.

The second step deals with the fact that firms could leave the path, given the path expectations. Consider the example of firm A's monopoly with blockaded entry of B. In the third stage, firm B selects the optimal quantity of good B, given the path expectations and firm A's optimal quantity. If it were beneficial for firm B to supply a positive quantity of good B, *i.e.* to leave the path, the path would not be consistent any longer and consumers would anticipate firm B to deviate. Thus, rational expectations exclude this path. For the other paths, we can define similar stability conditions.

4.3.1 Profit-Maximizing Quantities with Fulfilled Path Expectations

This section deals with the first step of computing a fulfilled expectations equilibrium. For a given path, we will derive the profit-maximizing quantities which are equal to the expected ones.

4.3.1.1 Monopoly with Blockaded Entry of the Competitor

Suppose that firm i has a considerable cost advantage so that the entry of technology j is blockaded. The profit function of monopolist i is given by

$$\Pi_{i,M} = y_{i,M}\left(\alpha_i - y_{i,M} + n\,y_i^e - c_i\right). \tag{4.2}$$

Maximizing the profit function with respect to $y_{i,M}$ results in the monopoly quantity for given expectations:

$$y_{i,M}{}^*(y_i^e) = \frac{\alpha_i - c_i + n\,y_i^e}{2}. \tag{4.3}$$

The maximum monopoly profit for given expectations is equal to

$$\Pi_{i,M}{}^*(y_i^e) = \frac{(\alpha_i - c_i + n\,y_i^e)^2}{4}. \tag{4.4}$$

The expected quantity y_i^e must be set equal to the actual quantity $y_{i,M}{}^*$. This defines the level of fulfilled expectations as the solution of

$$y_{i,M}^e{}^* = y_{i,M}{}^*(y_{i,M}^e{}^*) \quad \Leftrightarrow \quad y_{i,M}^e{}^* = \frac{\alpha_i - c_i}{2 - n}, \tag{4.5}$$

where $y_{i,M}{}^*(y_i^e)$ was substituted from Equation (4.3). The maximum monopoly profit with fulfilled expectations is then given by

$$\Pi_{i,M}^e{}^* = \frac{(\alpha_i - c_i)^2}{(n-2)^2}. \tag{4.6}$$

4.3.1.2 Incompatible Duopoly

In the second stage, firm A selects $y_{A,I}$ as the quantity of good A. Firm B responds in the third stage with quantity $y_{B,I}$ of product B. The profit functions of the duopolists are given by

$$\Pi_i = y_i(\alpha_i - y_i - \gamma y_j + n y_i^e - c_i). \tag{4.7}$$

Maximizing the follower's profit function with respect to $y_{B,I}$ results in the follower's optimal quantity for given expectations y_B^e and for the given leader quantity $y_{A,I}$:

$$y_{B,I}{}^*(y_B^e, y_{A,I}) = \frac{\alpha_B - c_B - \gamma y_{A,I} + n y_B^e}{2}. \tag{4.8}$$

Substituting Equation (4.8) in the profit function of firm A and maximizing with respect to $y_{A,I}$ yields the optimal leader quantity for given expectations y_A^e and y_B^e:

$$y_{A,I}{}^*(y_A^e, y_B^e) = \frac{2(\alpha_A - c_A) - \gamma(\alpha_B - c_B) - n(\gamma y_B^e - 2 y_A^e)}{2(2 - \gamma^2)}. \tag{4.9}$$

Substituting Equation (4.9) back in Equation (4.8) yields the follower's best reply

$$y_{B,I}{}^*(y_A^e, y_B^e) =$$
$$\frac{(4 - \gamma^2)(\alpha_B - c_B) - 2\gamma(\alpha_A - c_A) - n(\gamma^2 y_B^e - 4 y_B^e + 2\gamma y_A^e)}{4(2 - \gamma^2)}. \tag{4.10}$$

After having derived the optimal quantities of A and B as functions of the expected network sizes, we can endogenize consumers' expectations. For this purpose, we make use of Equations (4.9) and (4.10) to define the fixed point conditions $y_{i,I}^e* = y_{i,I}*(y_{i,I}^e*, y_{j,I}^e*)$, $i,j = A,B$, $i \neq j$. Solving these conditions for $y_{A,I}^e*$ and $y_{B,I}^e*$ results in the optimal quantities of A and B which fulfill expectations:

$$y_{A,I}^e* = \frac{2\gamma(\alpha_B - c_B) - (4 - 2n)(\alpha_A - c_A)}{(4-n)\gamma^2 - 2(n^2 - 4n + 4)},$$ (4.11)

$$y_{B,I}^e* = \frac{2\gamma(\alpha_A - c_A) + (\gamma^2 - 4 + 2n)(\alpha_B - c_B)}{(4-n)\gamma^2 - 2(n^2 - 4n + 4)}.$$ (4.12)

Maximum profits of firm A and B are given by:

$$\Pi_{A,I}^e* = \frac{2(2-\gamma^2)[\gamma(\alpha_B - c_B) - (2-n)(\alpha_A - c_A)]^2}{[2(n^2 - 4n + 4) - \gamma^2(4-n)]^2},$$ (4.13)

$$\Pi_{B,I}^e* = \frac{[2\gamma(\alpha_A - c_A) + (\gamma^2 + 2n - 4)(\alpha_B - c_B)]^2}{[2(n^2 - 4n + 4) - \gamma^2(4-n)]^2}.$$ (4.14)

Fig. 4.2 shows the profit-maximizing quantities (as dashed lines) and the maximum profits with fulfilled expectations. Quantities and profits are functions of alternative marginal costs c_B. The lower and the upper bounds of the incompatible duopoly are denoted by $c_{B,1}$ and $c_{B,2}$, respectively. At the equilibrium level, the quantities have to be positive, i.e. $y_{A,I}^e* > 0$ and $y_{B,I}^e* > 0$ must hold. After rearranging Equations (4.11) and (4.12), we see that an incompatible duopoly with fulfilled expectations exists if

$$c_{B,1} = \alpha_B - \frac{(2-n)(\alpha_A - c_A)}{\gamma} < c_B < c_{B,2} = \alpha_B - \frac{2\gamma(\alpha_A - c_A)}{4 - 2n - \gamma^2}.$$ (4.15)

In Fig. 4.2, the quantities and profits of firm A rise with c_B. On the other hand, the quantities and profits of firm B decrease with c_B. Generally, quantities and profits show normal reactions to cost variations if

$$\frac{\partial y_{i,I}^{e}\,*}{\partial c_{j}} = \frac{2\gamma}{2(n^{2}-4n+4)-(4-n)\gamma^{2}} > 0 \quad \text{with} \quad i,j = A,B, \ i \neq j,$$

$$\frac{\partial y_{B,I}^{e}\,*}{\partial c_{B}} = \frac{\gamma^{2}+2n-4}{2(n^{2}-4n+4)-(4-n)\gamma^{2}} < 0 \quad \text{and}$$

$$\frac{\partial y_{A,I}^{e}\,*}{\partial c_{A}} = \frac{2n-4}{2(n^{2}-4n+4)-(4-n)\gamma^{2}} < 0 \quad \text{holds.}$$

Since $2\gamma > 0$, $\gamma^{2}+2n-4<0$ and $2n-4<0$, the denominators must be positive. Making use of this condition, we have

$$\frac{\partial y_{i,I}^{e}\,*}{\partial c_{j}} > 0 \quad \text{and} \quad \frac{\partial y_{i,I}^{e}\,*}{\partial c_{i}} < 0 \quad \Leftrightarrow \quad n < n_{ID} = 2 - \frac{\gamma^{2}+\gamma\sqrt{\gamma^{2}+16}}{4}. \tag{4.16}$$

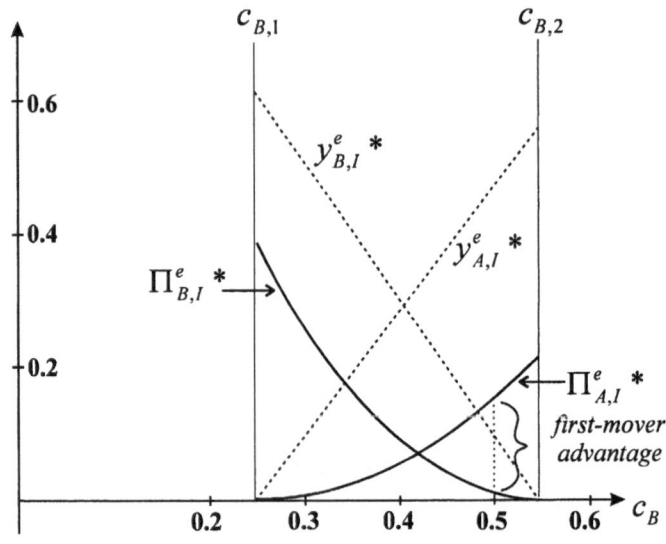

Fig. 4.2: Profit-maximizing quantities and maximum profits with fulfilled expectations for parameters $\alpha_{A} = \alpha_{B} = 1$, $n = \gamma = 0.8$, $c_{A} = 0.5$

In the following analysis, we will assume that $n < n_{ID}$ holds. With increasing network parameter n and rising degree of substitutability γ, the existence area $\left[c_{B,1}, c_{B,2}\right]$ shrinks and the functions $y_{A,I}^e *$ and $y_{B,I}^e *$ become steeper. The case of $c_{B,1} = c_{B,2}$ corresponds to the situation of $n = n_{ID}$: The incompatible duopoly does not exist any longer and $y_{A,I}^e *$ and $y_{B,I}^e *$ are vertical lines.

Proposition 4.1: *The incompatible duopoly occurs as a fulfilled expectations equilibrium if $c_{B,1} < c_B < c_{B,2}$ holds.*

4.3.1.3   Entry Deterrence

In the second stage, quantity leader A can supply an excessive quantity of good A in order to prevent the entry of technology B in the following stage. The optimal response of firm B, given by Equation (4.8), must be set equal to zero. Substituting rational expectations $y_B^e = 0$ and solving for y_A results in the deterrence output which fulfills expectations:

$$y_{B,I}^e * = \frac{\alpha_B - c_B - \gamma\, y_{A,D} \overset{!}{}}{2} = 0 \quad \Leftrightarrow \quad y_{A,D}^e * = \frac{\alpha_B - c_B}{\gamma}. \tag{4.17}$$

Substitution of $y_{A,D}^e *$ and $y_B = 0$ in the profit function of firm A, given by Equation (4.7), yields the deterrence profit with fulfilled expectations:

$$\Pi_{A,D}^e * = \frac{(\alpha_B - c_B)\left[\gamma(\alpha_A - c_A) - (1-n)(\alpha_B - c_B)\right]}{\gamma^2}. \tag{4.18}$$

Fig. 4.3 depicts the deterrence quantities and profits as functions of alternative marginal costs c_B. In the interval $x' \le c_B \le 1$, firm A's deterrence profit, $\Pi_{A,D}^e *$, rises with decreasing marginal costs, c_B, of its rival. To identify the reason for this perverse cost effect, it is necessary to scrutinize firm A's profit function. The deterrence profit is composed of the deterrence revenue, $r_{A,D}^e *$, and total costs $c_A\, y_{A,D}^e *$. As shown in Fig. 4.3, firm A has to supply a larger deterrence quantity, $y_{A,D}^e *$, with decreasing c_B. The larger network size results in an increased willingness to pay for good A.[40] The deterrence revenue is hyperbolic, it is maximal at x and goes back to zero because positive net-

---

[40] The entry deterrence strategy is not necessarily a credible commitment. Later on, we will derive the conditions for credibility.

work effects of the quantity expansion are finally offset by negative price effects. The deterrence profit is positive if

$$\Pi_{A,D}^e * > 0 \quad \Leftrightarrow \quad \alpha_B - \frac{\gamma(\alpha_A - c_A)}{1-n} < c_B < \alpha_B \text{ holds.} \qquad (4.19)$$

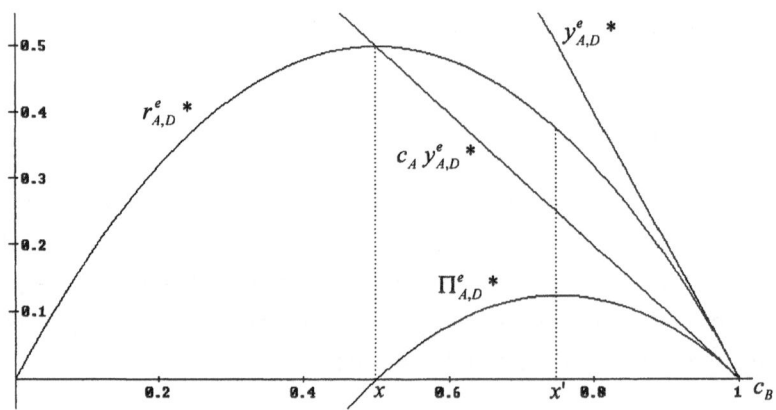

Fig. 4.3: Entry deterrence output and profit for parameters $\alpha_A = \alpha_B = 1$, $n = \gamma = c_A = 0.5$

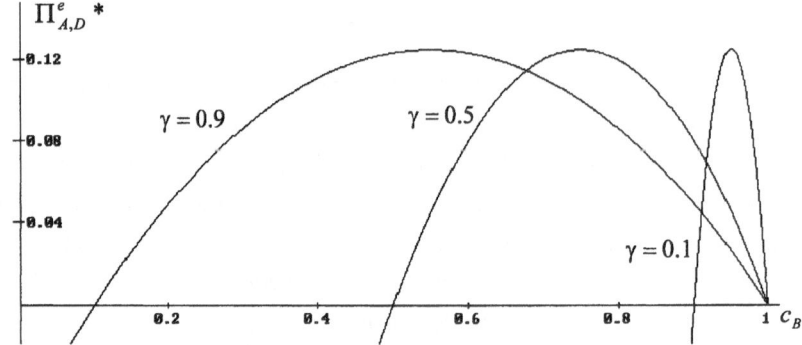

Fig. 4.4: Impact of substitutability on the deterrence profit for parameters $\alpha_A = \alpha_B = 1$, $n = c_A = 0.5$

Fig. 4.4 illustrates the impact of substitutability on deterrence profits. If the degree of substitutability, γ, is low, firm A needs a larger quantity in order to prevent the entry of technology B. Thus, starting with $c_B = 1$, the maximum deterrence profit $\Pi^e_{A,D} * (c_B^{max})$ and the zero point are reached faster.

Fig. 4.4 also demonstrates that the maximum level is independent of the degree of substitutability:

$$c_B^{max} = \frac{2(1-n)\alpha_B - \gamma(\alpha_A - c_A)}{2(1-n)} \; ; \; \Pi^e_{A,D} * (c_B^{max}) = \frac{(\alpha_A - c_A)^2}{4(1-n)}. \qquad (4.20)$$

The impact of the network parameter n on the deterrence profit is unambiguous, *i.e.* the deterrence profit always rises in n:

$$\frac{d\Pi^e_{A,D} *}{dn} = \frac{(\alpha_B - c_B)^2}{\gamma^2} > 0. \qquad (4.21)$$

Fig. 4.5 depicts this result. If n approaches its upper limit $n = 1$, the deterrence profit is illustrated by a straight line.

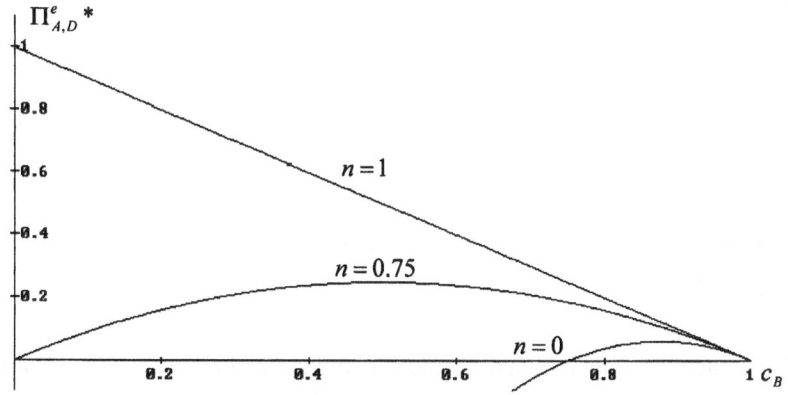

Fig. 4.5: Impact of network effects on the deterrence profit for parameters $\alpha_A = \alpha_B = 1$, $\gamma = c_A = 0.5$

4.3.2   Consistency of Equilibrium Paths

So far, we have derived the profit-maximizing quantities which fulfill expectations on a given path. For example, in the case of blockaded entry, it was assumed that it is not beneficial for the weaker firm to enter the market. The second step of the equilibrium analysis deals with the fact that firms may deviate from the path.

4.3.2.1 Blockaded Entry of Technology B

Suppose that the entry of technology B is blockaded. The possible equilibrium path with fulfilled expectations is given by:

First stage: Consumers form the expectations $y_A^e = y_{A,M}^e *$ and $y_B^e = 0$.

Second stage: Firm A selects the monopoly quantity $y_{A,M}^e *$.

Third stage: Firm B does not enter the market, i.e. $y_{B,I}^e * = 0$.

Starting with the third stage, we examine whether it is beneficial for firm B to choose $y_{B,I}^e * = 0$. Substituting path expectations $y_B^e = 0$ and the leader quantity $y_A = y_{A,M}^e *$ in firm B's best reply, given by Equation (4.8), yields:

$$\bar{y}_B = y_B(y_B^e = 0, y_A = y_{A,M}^e *) = \frac{(2-n)(\alpha_B - c_B) - \gamma(\alpha_A - c_A)}{2(2-n)}. \qquad (4.22)$$

$$\bar{y}_B \le 0 \quad \Leftrightarrow \quad c_B \ge c_{B,3} = \alpha_B - \frac{\gamma(\alpha_A - c_A)}{2-n}. \qquad (4.23)$$

Proposition 4.2: *The market entry of technology B is blockaded if $c_B \ge c_{B,3}$ holds.*

Note that the upper bound of the incompatible duopoly, $c_{B,2}$, is always smaller than the lower bound, $c_{B,3}$, of technology B's blockaded entry:

$$c_{B,2} = \alpha_B - \frac{2\gamma(\alpha_A - c_A)}{4 - 2n - \gamma^2} < c_{B,3} = \alpha_B - \frac{\gamma(\alpha_A - c_A)}{2-n}.$$

Rearranging the inequation yields

$$\frac{2\gamma(\alpha_A - c_A)}{4 - 2n} < \frac{2\gamma(\alpha_A - c_A)}{4 - 2n - \gamma^2},$$

which is always fulfilled. Hence, the blockaded entry of technology B excludes the incompatible duopoly.

4.3.2.2   Blockaded Entry of Technology A

The next step is to analyze under which circumstances the entry of technology A is blockaded. The possible equilibrium path is given by:

First stage: Consumers form the expectations $y_A^e = 0$ and $y_B^e = y_{B,M}^e$ *.

Second stage: Firm A does not enter the market, i.e. $y_A* = 0$.

Third stage: Firm B supplies the monopoly quantity $y_{B,M}^e$ *.

In order to examine whether it is beneficial for firm A to choose $y_A* = 0$, we have to derive firm B's optimal reply to the leader quantity for given path expectations $y_B^e = y_{B,M}^e$ *:

$$y_B(y_A) = \frac{2(\alpha_B - c_B) - (2 - n)\gamma y_A}{2(2 - n)}. \tag{4.24}$$

Substituting Equation (4.24) and $y_A^e = 0$ in firm A's profit function and maximizing with respect to y_A results in firm A's optimal quantity in the case where firm A supplies a positive quantity - i.e. it leaves the path - in the second stage:

$$\bar{y}_A = \frac{\gamma(\alpha_B - c_B) - (2 - n)(\alpha_A - c_A)}{(2 - n)(\gamma^2 - 2)}, \tag{4.25}$$

$$\bar{y}_A \leq 0 \quad \Leftrightarrow \quad c_B \leq c_{B,1} = \alpha_B - \frac{(2 - n)(\alpha_A - c_A)}{\gamma}. \tag{4.26}$$

Proposition 4.3: *The market entry of technology A is blockaded if* $c_B \leq c_{B,1}$ *is given.*

Note that the upper bound of technology A's blockaded entry is equal to the lower bound of the incompatible duopoly. Hence, the incompatible duopoly cannot occur as a fulfilled expectations equilibrium if the entry of technology A is blockaded.

4.3.2.3   Entry Deterrence

Firm A can exploit its first-mover advantage in order to prevent the entry of technology B. But the deterrence path is only consistent if it is worthwhile for firm A to select the deterrence output $y^e_{A,D}$ * in the second stage, given the path expectations. The possible equilibrium path is given by:

First stage: Consumers form the expectations $y^e_A = y^e_{A,D}$ * and $y^e_B = 0$.

Second stage: Firm A supplies the deterrence quantity $y^e_{A,D}$ *.

Third stage: Firm B does not enter the market, $i.e.$ $y_B = 0$.

Suppose that firm A selects a lower quantity $y_A < y^e_{A,D}$ * in the second stage. In this case, it is beneficial for firm B to enter the market in the third stage. Its best reply for given expectations $y^e_B = 0$ is equal to:

$$y_B *(y_A) = \frac{\alpha_B - c_B - \gamma\, y_A}{2} > 0. \tag{4.27}$$

Substituting $y_B *(y_A)$ and $y^e_A = y^e_{A,D}$ * in firm A's profit function yields:

$$\Pi_{A,I}(y_A)$$
$$= \frac{y_A(2\gamma\alpha_A + 2n\alpha_B - \gamma^2\alpha_B - 2\gamma c_A + \gamma^2 c_B - 2nc_B)}{2\gamma} - \frac{(y_A)^2(2-\gamma^2)}{2}. \tag{4.28}$$

Maximizing Equation (4.28) with respect to y_A and substituting the deterrence quantity $y^e_{A,D}$ * results in:

$$\frac{d\Pi_{A,I}}{d y_A} = \frac{2\gamma(\alpha_A - c_A) + (\gamma^2 + 2n - 4)(\alpha_B - c_B)}{2\gamma}. \tag{4.29}$$

Entry deterrence is not consistent if it is worthwhile for the leader to reduce its quantity y_A and to accept firm B's market entry. In this case, the first derivative of firm A's profit function must be negative:

$$\frac{d\Pi_{A,I}}{dy_A} < 0 \quad \Leftrightarrow \quad c_B < c_{B,2}. \tag{4.30}$$

Note that $c_{B,2}$ is the upper bound of the incompatible duopoly. Given the deterrence output $y_{A,D}^e *$ and the deterrence expectations $y_A^e = y_{A,D}^e *$ and $y_B^e = 0$, firm A could increase its profit by reducing its output. Thus, entry deterrence is not consistent in this case. Rational consumers would anticipate the deviation from the deterrence path for values lower than $c_{B,2}$. However, if $c_B > c_{B,2}$ holds, firm A would always lower its profit by reducing the output. This is a necessary, but not sufficient, condition for the consistency of entry deterrence.

The next step is to analyze under which circumstances firm A would deviate from the deterrence path by *increasing* its output. Suppose that firm A chooses $y_A > y_{A,D}^e *$ in the second stage. In this case, firm B would not enter the market in the third stage. Firm A's monopoly profit for given expectations $y_A^e = y_{A,D}^e *$ and $y_B^e = 0$ and for the actual quantity $y_B = 0$ is equal to

$$\Pi_{A,M}(y_A) = \frac{y_A[\gamma(\alpha_A - c_A) + n(\alpha_B - c_B)]}{\gamma} - (y_A)^2. \tag{4.31}$$

Maximizing Equation (4.31) with respect to y_A and substituting the deterrence quantity $y_A = y_{A,D}^e *$ results in:

$$\frac{d\Pi_{A,M}}{dy_A} = \alpha_A - c_A - \frac{(2-n)(\alpha_B - c_B)}{\gamma}. \tag{4.32}$$

Suppose that the first derivative of firm A's profit function is positive:

$$\frac{d\Pi_{A,M}}{dy_A} > 0 \quad \Leftrightarrow \quad c_B > c_{B,3}. \tag{4.33}$$

In this case, firm A would select a quantity which exceeds the deterrence output. Hence, the first-stage expectations $y_A^e = y_{A,D}^e *$ are not consistent with firm A's behavior in the second stage. Thus, if technology B's entry is blockaded, $y_A^e = y_{A,M}^e *$ is the only equilibrium with fulfilled expectations.

Proposition 4.4: *Entry deterrence is a fulfilled expectations equilibrium if and only if $c_{B,2} < c_B < c_{B,3}$ holds, i.e. the incompatible duopoly and the monopoly with blockaded entry of the competing technology are not feasible.*

Proposition 4.5: *Let $c'_{B,3}$ be infinitesimally smaller than $c_{B,3}$. If the network effect is positive, $n > 0$, we have: $\Pi^e_{A,D} * (c'_{B,3}) > \Pi^e_{A,D} * (c_{B,3}) = \Pi^e_{A,M} *$. For a sufficiently strong network effect $n > n_{ED}$, $\Pi^e_{A,D} *$ exceeds $\Pi^e_{A,M} *$ throughout the relevant interval $c_{B,2} < c_B < c_{B,3}$.*

Proof: We have that $\Pi^e_{A,D} * (c_{B,3}) = \Pi^e_{A,M} *$ is always fulfilled.
The first derivative of $\Pi^e_{A,D} *$ is negative at $c_{B,3}$:

$$\frac{\partial \ \Pi^e_{A,D} * (c_{B,3})}{\partial c_B} = -\frac{n(\alpha_A - c_A)}{\gamma(2-n)} < 0 \ . \tag{4.34}$$

The limit n_{ED} follows from $\Pi^e_{A,D} * (c_{B,2}) > \Pi^e_{A,M} *$:

$$n > n_{ED} = 1 - \frac{\sqrt{2}\sqrt{(2-\gamma^2)}}{2} \ . \tag{4.35}$$

Since $\Pi^e_{A,D} * (c_B)$ is concave with

$$\frac{\partial^2 \ \Pi^e_{A,D} *}{\partial c_B^{\ 2}} = -\frac{2(1-n)}{\gamma^2} < 0$$

and continuous,

$$\Pi^e_{A,D} * (c_B) > \Pi^e_{A,M} * \text{ holds}$$

throughout interval $c_{B,2} < c_B < c_{B,3}$. $\hspace{2cm}$ *QED*

4.3.3   Numerical Example

Fig. 4.6 depicts the different fulfilled expectations equilibria for a numerical example. The deterrence profit, $\Pi^e_{A,D} *$, rises with decreasing marginal costs of technology B because the incumbent firm can credibly commit to a larger network. Since the network effects are sufficiently strong, the deterrence profit is always higher than firm A's monopoly profit with blockaded entry of technology B.

Fig. 4.7 shows the equilibrium areas, where the marginal costs of both technologies are variable. Note that Fig. 4.6 is a cross-section of Fig. 4.7 at $c_A = 0.5$. The figure illustrates that the deterrence strategy is only feasible if firm A has a moderate cost advantage.

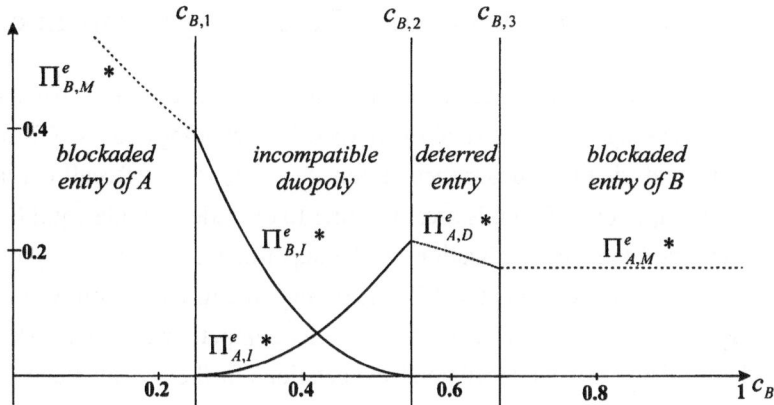

Fig. 4.6: Fulfilled expectations equilibria for parameters $\alpha_A = \alpha_B = 1$, $n = \gamma = 0.8$, $c_A = 0.5$

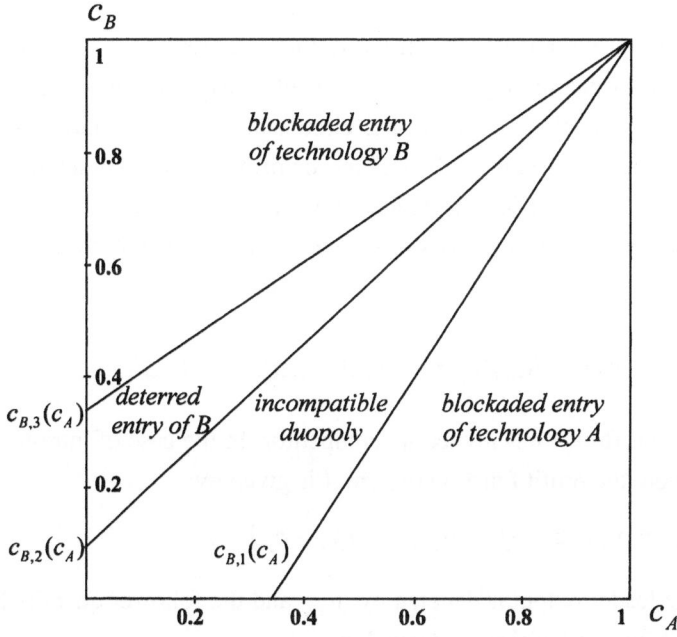

Fig. 4.7: Two-dimensional representation of fulfilled expectations equilibria for parameters $\alpha_A = \alpha_B = 1$, $n = \gamma = 0.8$

4.4   Entry Deterrence vs. Intra-Technology Competition

In this section, we will analyze the incumbent's strategic choice of whether to share its technology with the rival or to keep its technology for itself. Let us restrict the analysis to the deterrence region $c_{B,2} < c_B < c_{B,3}$. Recall that technology A is superior to B in this area. In order to exclude multiple equilibria, it is also assumed that standardization on the superior technology A (by means of deterrence or invitation) is focal.[41] The timing structure is as follows: In the first stage, firm A decides whether to invite its competitor into network A or not. This decision affects consumers' expectations in the second stage. After having invited its competitor, firm A selects its leader quantity in the third stage. Finally, firm B chooses the optimal follower quantity of good A in the fourth stage. It is assumed that firm B can either join network A or enter the market with its own technology B but not use both strategies at the same time. Moreover, suppose that both firms have the same marginal costs c_A if they produce good A.

The equilibrium analysis is similar to the preceding one. First, we will derive the profit-maximizing quantities that fulfill expectations on a given path. Then, we will analyze under which circumstances the invitation strategy is self-enforcing, $i.e.$ it must be beneficial for firm B to join voluntarily network A instead of selecting its own technology B. The last step of the equilibrium analysis deals with firm A's first-stage decision, $i.e.$ whether to invite its competitor to engage in intra-technology competition or not.

4.4.1   Profit-Maximizing Quantities with Fulfilled Expectations

Suppose that firm A has invited its competitor. In the case of intra-technology competition, the profit function of firm A is given by:

$$\Pi_{A,L} = y_{A,L} \left(\alpha_A - y_{A,L} - y_{A,F} + n y_A^e - c_A \right) \quad , \qquad (4.36)$$

where the leader output is denoted by $y_{A,L}$ and the follower quantity by $y_{A,F}$. The profit function of firm B is equal to

---

[41] Therefore, it is not necessary to analyze the invitation of firm A into network B. Suppose that both firms invite each other into their networks. This coordination game, which corresponds to the "Battle of the Sexes", has multiple expectations and equilibria.

$$\Pi_{A,F} = y_{A,F}\left(\alpha_A - y_{A,L} - y_{A,F} + n y_A^e - c_A\right). \tag{4.37}$$

Optimal quantities for given expectations are equal to:

$$y_{A,L}*(y_A^e) = \frac{\alpha_A - c_A + n y_A^e}{2}, \tag{4.38}$$

$$y_{A,F}*(y_A^e) = \frac{\alpha_A - c_A + n y_A^e}{4}. \tag{4.39}$$

Summing up yields the total output for given expectations:

$$y_A*(y_A^e) = \frac{3(\alpha_A - c_A + n y_A^e)}{4}. \tag{4.40}$$

Maximum profits for given expectations are equal to:

$$\Pi_{A,L}*(y_A^e) = \frac{(\alpha_A - c_A + n y_A^e)^2}{8}, \tag{4.41}$$

$$\Pi_{A,F}*(y_A^e) = \frac{(\alpha_A - c_A + n y_A^e)^2}{16}. \tag{4.42}$$

The fixed point condition is derived from Equation (4.40):

$$y_A^e* = y_A*(y_A^e*) \quad \Leftrightarrow \quad y_A^e* = \frac{3(\alpha_A - c_A)}{4 - 3n}. \tag{4.43}$$

Optimal quantities and profits with fulfilled expectations are given by Equations (4.44)-(4.47):

$$y_{A,L}^e* = \frac{2(\alpha_A - c_A)}{4 - 3n}, \tag{4.44}$$

$$y_{A,F}^e* = \frac{\alpha_A - c_A}{4 - 3n}, \tag{4.45}$$

$$\Pi_{A,L}^e* = \frac{2(\alpha_A - c_A)^2}{(3n - 4)^2}, \tag{4.46}$$

$$\Pi^e_{A,F}* = \frac{(\alpha_A - c_A)^2}{(3n-4)^2}. \tag{4.47}$$

Note that the leader profit is twice as high as the follower profit. This result reflects the first mover advantage of the Stackelberg leader.

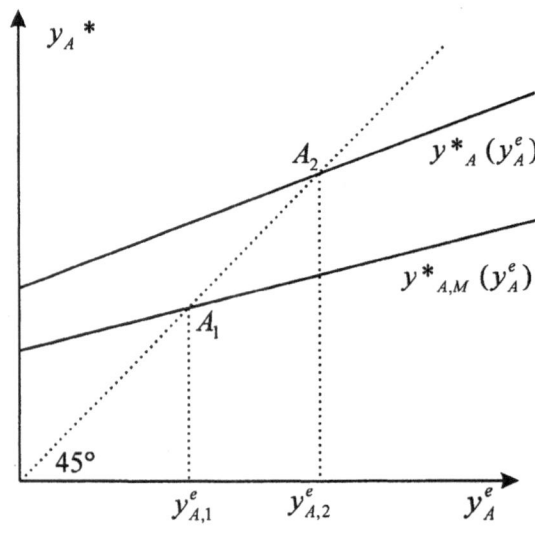

Fig. 4.8: Impact of the invitation on the expected network size

Fig. 4.8 illustrates the impact of firm B's entry on the expected network size of good A. Suppose that the entry of technology B is blockaded. In the monopoly case, A_1 represents the optimal quantity of good A which fulfills expectations. This quantity is given by the intersection point of $y*_{A,M}(y^e_A)$ with the 45°-line. By means of firm B's entry into network A, the curve $y*_A(y^e_A)$ is shifted outward. The optimal quantity which fulfills expectations increases to A_2. The invitation is the only way to commit credibly to a network size which exceeds firm A's profit-maximizing output. If firm A promised to produce the output A_2 by itself, it would not be credible because, as a monopolist, firm A has an incentive to reduce output for any given level of expectations exceeding A_1. At $y^e_A = 0$, the curve $y_A*(y^e_A)$ is above the 45°-line because of $\alpha_A > c_A$. If its slope is smaller than one, the function has an intersection with the 45°-line:

$$\frac{d y_A *(y_A^e)}{d y_A^e} < 1 \quad \Leftrightarrow \quad n < \frac{4}{3},$$ (4.48)

which is always fulfilled because of the assumption $n < 1$. If firm A is a monopolist, the existence condition is relaxed. From Equation (4.3) follows that $n < 2$ must hold in this case.

4.4.2   Consistency of the Invitation Strategy

So far, we have derived the profit-maximizing quantities which fulfill expectations on a given path. Thus, in the case of intra-technology competition, it was assumed that it is worthwhile for firm B to join network A. Now, it must be taken into account that firm B could deviate from this path. Duopoly within market A is given by the following path:

First stage: Firm A invites its competitor into network A.

Second stage: Consumers form the expectations $y_A^e = y_{A,L}^e * + y_{A,F}^e *$ and $y_B^e = 0$.

Third stage: Firm A selects the leader quantity $y_{A,L}^e *$ of good A.

Fourth stage: Firm B reacts with the follower quantity $y_{A,F}^e *$ of good A.

Starting with the last stage, we analyze whether it is really beneficial for firm B to accept the invitation into network A. First, the profit-maximizing quantity of good B must be derived, given the leader output $y_{A,L}^e *$ and the expectations $y_B^e = 0$. Substituting $y_{A,L}^e *$ and $y_B^e = 0$ in firm B's best reply, given by Equation (4.8), yields:

$$\hat{y}_B = y_B(y_B^e = 0, y_A = y_{A,L}^e *) = \frac{(4-3n)(\alpha_B - c_B) - 2\gamma(\alpha_A - c_A)}{2(4-3n)}.$$ (4.49)

Firm B's maximum profit for deviating from the invitation path is given by:

$$\hat{\Pi}_B = \begin{cases} \dfrac{[2\gamma(\alpha_A - c_A) - (4-3n)(\alpha_B - c_B)]^2}{4(3n-4)^2} & , if \ \hat{y}_B > 0 \\ \\ 0 & , otherwise \end{cases}.$$ (4.50)

Firm B voluntarily joins network A if the follower profit within network A exceeds the deviation profit $\hat{\Pi}_B$:

$$\Pi^e_{A,F} * > \hat{\Pi}_B \quad \Leftrightarrow \quad c_B > \hat{c}_B = \alpha_B - \frac{2(\alpha_A - c_A)(1 + \gamma)}{4 - 3n} . \tag{4.51}$$

Proposition 4.6: *Firm B accepts the invitation into network A if $c_B > \hat{c}_B$ holds. Otherwise, the invitation path cannot be a fulfilled expectations equilibrium because firm B would run its own technology B.*

Fig. 4.9 illustrates the consistency condition for the invitation path. For $c_B < \hat{c}_B$, firm B's deviation profit $\hat{\Pi}_B$ exceeds its follower profit $\Pi^e_{A,F} *$ within market A. Thus, the invitation strategy is not feasible. If $c_B > \hat{c}_B$ is given, the invitation strategy is self-enforcing.

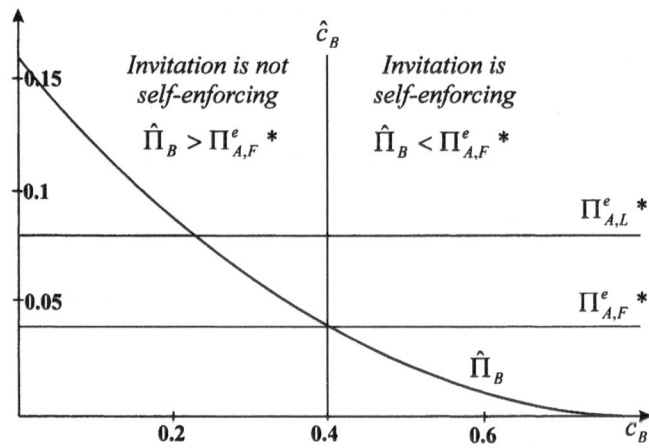

Fig. 4.9: Consistency of the invitation strategy for parameters $\alpha_A = \alpha_B = 1$, $n = \gamma = c_A = 0.5$

4.4.3   The Invitation Decision

It remains to be analyzed under which circumstances firm A will invite its competitor instead of selecting the deterrence output of good A. If the invitation strategy is feasible, firm A has to compare its invitation payoff with its deterrence payoff.

Proposition 4.7: *Suppose that entry deterrence is feasible, i.e.* $c_{B,2} < c_B < c_{B,3}$. *Network A is closed and entry deterrence occurs as a fulfilled expectations equilibrium if* $\Pi_{A,D}^e{}^* > \Pi_{A,L}^e{}^*$ *holds. Conversely, if* $\Pi_{A,D}^e{}^* < \Pi_{A,L}^e{}^*$ *is given, the invitation of firm B is a fulfilled expectations equilibrium.*[42]

Fig. 4.10 compares the deterrence profits with the profits in the case of intra-technology competition. If $\tilde{c}_B < c_B$ holds, firm A invites its competitor. For $c_{B,2} < c_B < \tilde{c}_B$, firm A selects the deterrence strategy. Thus, in this situation deterrence is a more successful commitment device than the invitation strategy.

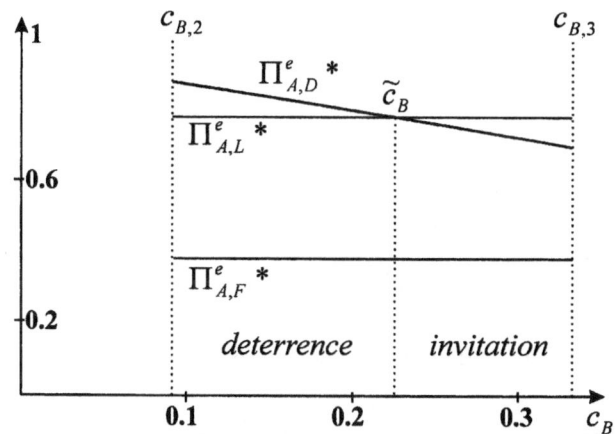

Fig. 4.10: Invitation into network A vs. entry deterrence for parameters $\alpha_A = \alpha_B = 1$, $n = \gamma = 0.8$ and $c_A = 0$

---

[42] Since $\hat{c}_B < c_{B,2}$ always holds, the invitation into network A is self-enforcing throughout the deterrence interval $c_{B,2} < c_B < c_{B,3}$.

4.4.4   Welfare Implications

Fig. 4.11 illustrates the demand curves with given and with fulfilled expectations. After connecting all the points with identical expected and real sales, we can derive the demand curve with fulfilled expectations, which is flatter than the demand curves with given expectations. Consumers' surplus must be based on rational expectations. If the real quantity of good A is reduced, the willingness to pay increases by a lower amount because consumers face a smaller network. Thus, consumers' surplus corresponds to the dotted area below the demand curve with fulfilled expectations.

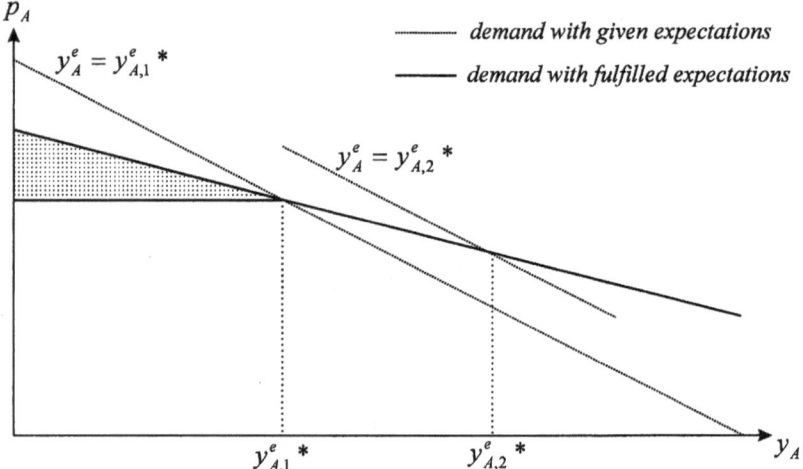

Fig. 4.11: Consumers' surplus and given expectations

For a single market i, consumers' surplus is given by:

$$CS_i^e * = \frac{(1-n)(y_i^e *)^2}{2}.$$

(4.52)

Substituting Equation (4.43) into Equation (4.52) yields consumers' surplus in the case of duopoly in market A:

$$CS_{A,C}^e * = \frac{9(1-n)(\alpha_A - c_A)^2}{2(3n-4)^2}.$$

(4.53)

Substituting Equation (4.5) into Equation (4.52) leads to consumers' surplus in the case of monopoly:

$$CS_{i,M}^{e} * = \frac{(1-n)(\alpha_i - c_i)^2}{2(n-2)^2}.$$ (4.54)

For entry deterrence, Equation (4.17) must be substituted into Equation (4.52):

$$CS_{A,D}^{e} * = \frac{(1-n)(\alpha_B - c_B)^2}{2\gamma^2}.$$ (4.55)

A standard welfare measure can be derived by adding up consumers' surplus and total profits. Substituting Equations (4.46), (4.47) and (4.53) yields welfare in the case of duopoly in market A:

$$W_{A,C}^{e} * = CS_{A,C}^{e} * + \Pi_{A,L}^{e} * + \Pi_{A,F}^{e} * = \frac{3(5-3n)(\alpha_A - c_A)^2}{2(3n-4)^2}.$$ (4.56)

In the monopoly case, welfare corresponds to

$$W_{i,M}^{e} * = CS_{i,M}^{e} * + \Pi_{i,M}^{e} * = \frac{(3-n)(\alpha_i - c_i)^2}{2(n-2)^2}.$$ (4.57)

Finally, in the deterrence case welfare is given by

$$W_{A,D}^{e} * = CS_{A,D}^{e} * + \Pi_{A,D}^{e} *$$
$$= \frac{(\alpha_B - c_B)(2\gamma(\alpha_A - c_A) - (1-n)(\alpha_B - c_B))}{2\gamma^2}.$$ (4.58)

Proposition 4.8: *Throughout the interval $c_{B,2} < c_B < c_{B,3}$, $CS_{A,C}^{e} * > CS_{A,D}^{e} *$ holds. Thus, consumers always prefer duopoly in market A to entry deterrence. Considering total welfare, we see that $W_{A,C}^{e} * > W_{A,D}^{e} *$ holds throughout the deterrence region. Thus, intra-technology competition is desirable from the welfare perspective.*

Proof: See appendix A 2.

4.5 Concluding Remarks

In this model, we have formalized the problem of whether an incumbent firm
has an incentive to share its technology with a rival firm or to keep its technology for itself. By sharing its technology, the incumbent credibly commits
to a future network size which exceeds the profit-maximizing monopoly
quantity. As a consequence, the incumbent reaps the benefits of the consumers' increased willingness to pay for the network good. What drives the model
is the assumption that consumers form rational expectations about the network
size before the firms are able to choose their output levels.[43] This timing
structure (including the Stackelberg assumption) is the same as in the model
by Economides (1996a). Whereas Economides' model is confined to intra-
technology competition, we have presented a framework dealing both with
inter-technology and intra-technology competition.

We have analyzed the problem in two steps. The first step was restricted to
inter-technology competition, *i.e.* the incumbent was assumed to keep the own
technology for itself. We have seen that the fulfilled expectations equilibria
depend on the relative marginal costs of both technologies. The entry of the
follower may be blockaded if the incumbent firm has strong cost advantages.
But if the follower has a considerable cost advantage, the incumbent's entry
can be blockaded as well. On the other hand, in the case of weak cost differences, both technologies coexist in an incompatible and heterogeneous duopoly. Entry deterrence is feasible only if the incumbent has a moderate cost
advantage. Each of the four possible fulfilled expectations equilibria has to
meet consistency conditions which ensure the uniqueness of the equilibrium.
The above results reflect the idea that markets are "tippy", *i.e.* the coexistence
of incompatible products may be unstable if firms are dissimilar in terms of
costs. Moreover, the region of the incompatible duopoly shrinks with strong
network effects and with a rising degree of substitutability.

The central argument of the model is that the incumbent may realize a
higher profit in the case of entry deterrence than in the situation where the

---

[43] The model deals with rational expectations in that consumers' expectations involve
all information available. Of course, in real market situations, there can be uncertainty (perhaps with respect to the pace of technological progress). See Farrell and
Katz (1998) for an analysis covering other types of expectations, most notably
"stubborn expectations".

competitor's entry is blockaded. The deterrence quantity not only prevents the competitor's entry, it also increases the incumbent's network and thus consumers' willingness to pay. Furthermore, we have derived the perverse effect that the deterrence profit can rise with decreasing marginal costs of the follower because the incumbent has to supply a larger deterrence quantity with increasing strength of its rival.

In the second part of the model, we have analyzed the incumbent's choice of whether to share its technology or to insist on inter-technology competition. We have seen that the deterrence strategy can be more profitable for the incumbent (if the incumbent's cost advantage is not too strong) than the strategy of sharing its technology. Since the incumbent does not internalize the positive externalities of the invitation strategy, which exists in terms of the follower profit and in an increased consumers' surplus, the deterrence strategy is welfare-inferior. Consequently, a policy recommendation should be to oblige the incumbent firm to license its technology to the rival.

5

Standardization of Nascent Technologies: The Tradeoff between Early Standardization and Experimentation

This chapter investigates the tradeoff between early (ex-ante) standardization and experimentation. The advantage of experimentation over ex-ante standardization is that users learn about the actual values of potential technologies ("learning by using") so that they can choose the ex-post standard with better information. However, experimentation involves a transient or even persistent loss of compatibility. We will numerically analyze the case where the values of two potential technologies are drawn from a bivariate normal distribution. The numerical analysis demonstrates that consumers prefer ex-ante standardization to experimentation if they expect the values of both technologies to be strongly correlated. Furthermore, the model shows that if the technologies are not equally attractive ex ante, there can be too much ex-ante standardization compared with the social optimum, or consumers may choose an inferior technology as ex-ante standard.

5.1 Introduction

A common characteristic of nascent technologies is that consumers cannot completely assess the product's quality at the time of market launch. The potentials and disadvantages of such technologies can be revealed only after consumers have experimented with them. In the course of being used, consumers may learn about new ways of utilization and about unexpected obstacles. Thus, consumers gain from variety by discovering the true values of competing technologies ("learning by using"). However, in the case of network technologies, variety also involves a transient or even persistent loss of compatibility. Early standardization, *i.e.* the adoption of a joint technology, may ensure compatibility from the beginning so that consumers benefit from positive network externalities.[44] However, early standardization also implies that the consumers do not find out the true qualities of alternative technologies which might be superior ex post.[45]

The market of DVD writers gives an example for experimentation with incompatible network technologies such as DVD-R, DVD-RAM, DVD-RW, DVD+RW, and DVD+R. In general, the distinctions rely on how the data is written to and read from the disk. For example, DVD-R and DVD+R disks can only be recorded once. Both DVD-R and DVD+R discs will play in most DVD-players, even older ones. DVD+RW, DVD-RW and DVD-RAM disks can all be recorded many of times. DVD-RAM, for instance, was created for storage of computer data like hard drive backups so that most DVD-players cannot play DVD-RAM discs. The other formats are better suited for recording movies, but they will only be compatible with newer DVD players.

In this chapter, we follow Choi (1996) who has developed a framework to analyze the tradeoff between early standardization and experimentation. The coordination problem of standardization is analyzed from the perspective of consumers.[46] For simplicity, the model considers two consumers and two competing network technologies. In the first period, the consumers choose between ex-ante standardization and experimentation. Experimentation means that the consumers adopt incompatible technologies while ex-ante standardization involves the adoption of a joint technology so that consumers benefit from positive network externalities. The first-period decision is based on limited information, *i.e.* the consumers only know the probability distribution of

---

[44] Alternatively, compatibility can be achieved by using converters. See, for example, Farrell and Saloner (1992).

[45] Note that variety has an entirely different role in this framework than in the model by Farrell and Saloner (1986b). In this early standardization model, Farrell and Saloner analyze the tradeoff between standardization and variety. Consumers are assumed to have heterogeneous preferences with respect to the good specifications. Thus, standardization entails the cost of constrained product variety.

[46] There exist several models analyzing the technology-adoption process. See, for instance, Farrell and Saloner (1986b), Belleflamme (1998), and Choi and Thum (1998).

both technologies' stand-alone values, which is "common knowledge".[47] In the second period, the values of all technologies used in the previous period become public knowledge. Based on this information, each consumer decides whether to stick to the own technology or to switch to the other one. We assume that switching entails switching costs.[48] Thus, the second-period game may result in persistence of incompatibility or in ex-post standardization.

The model distinguishes between two different regimes of coordination. The distinctive feature is whether consumers have vested interests or not. In the absence of vested interests, there exists a simple coordination problem which can be solved by "cheap talk".[49] We will interpret this situation as *committee standardization, i.e.* consumers convene in a standard-setting committee and communication is sufficient to ensure consensus. However, if consumers have vested interests, they face a coordination problem corresponding to the Battle of the Sexes. In this case, both consumers prefer a joint technology but they disagree about which technology to adopt. We will interpret this non-cooperative game as *de-facto standardization.*

In this chapter, we substantiate the outlined model framework by deriving numerical solutions for the case where the technologies' values are drawn from a bivariate normal distribution function.[50] This example of a continuous distribution function will enable us to investigate the impact of correlation on the incentives to experiment. It is shown that consumers prefer ex-ante standardization to experimentation if they expect the values of both technologies to be strongly correlated. This result is in contrast to Choi (p. 285f.) who "expect(s) experimentation to be a better option if the values of the two technologies tend to be negatively correlated." He confirms this intuition for a two-point distribution but he also acknowledges that "...the two-point distributions

---

[47] Here, the term "stand-alone value" denotes one user's basic utility in the case where the other user do not adopt the same technology. If both users adopt the same technology, each user also benefits from positive network externalities.

[48] See Marinoso (2001) for a model investigating endogenous switching costs. For an approach to estimate switching costs, see Shy (2002).

[49] For coordination by "cheap talk", see, for example, Farrell (1987) and Cooper et al. (1992).

[50] Choi assumes a symmetric probability distribution function without specifying the type. The comparative statics (p. 285) are not explicit enough because the network effect and the switching cost affect the integration regions and thus the probabilities of occurrence. To consider this effect, the probability distribution must be specified.

are not rich enough to capture the main point of the paper, which explains the use of a continuous distribution in the previous section" (Choi, p. 286, footnote 19).

The different impact of correlation on the results of the model can be traced back to Choi's assumption that the standard will not change in the second period if both users adopt the same technology in the first period. Nevertheless, Choi (p. 279, footnote 12) affirms that "this assumption is not crucial to the main results of the paper and can be easily dispensed with." We will dispense with this assumption, *i.e.* consumers can switch to the competing technology even if they have adopted the same technology in the first period. In this situation, consumers find out the true value of the chosen technology, only. But in the case of strong positive or negative correlation between both technologies, the consumers may use the observed value in order to learn something about the other value, *i.e.* they revise the expected ex-ante value of the other technology according to the Bayesian rule. Thus, experimentation only has a slight information advantage over ex-ante standardization in the case of strong correlation so that the information advantage of experimentation is easily outweighed by the compatibility advantage of ex-ante standardization.

Choi confines his model to the case where two potential technologies are equally attractive ex ante. Because of this symmetry, consumers have no vested interests in the first period so that they can coordinate their strategies in a standard-setting committee. However, by choosing different technologies in the first period, consumers build up vested interests. In the second period, there is then a coordination problem corresponding to the Battle of the Sexes. Due to switching costs, each consumer wants the other to join his technology but prefers to join the other's if the alternative is incompatibility. Finally, there may be too little ex-post standardization compared with the social optimum.

This chapter shows that the coordination problem aggravates if the technologies are not equally attractive ex ante, *i.e.* the first-period problem which technology to adopt may lead to inefficient results as well. In a numerical example, we will consider a risky technology with a low mean value and another technology with low deviation and a large mean value. It will be demonstrated that there is too little experimentation compared with the social optimum. Even if experimentation is worthwhile collectively, no consumer is willing to experiment with the risky technology due to its low mean value. In the end, there can be excessive ex-ante standardization on the technology with the larger mean value. Moreover, a coordination failure may occur in that the wrong (inferior) technology may become ex-ante standard.

The structure of this chapter is as follows: Section 5.2 deals with the model assumptions. In Section 5.3, the expected values of experimentation are derived. In Section 5.4, we will compute the expected values of ex-ante standardization. Section 5.5 deals with the first-period problem of whether to experiment with different technologies or to adopt the same technology. Equilibria and welfare implications are analyzed numerically, both for the case of a symmetric and an asymmetric probability distribution. Concluding remarks follow in Section 5.6.

5.2 Model Structure

Suppose that two homogeneous users choose between the incompatible and competing technologies A and B. The stand-alone value of technology A is denoted by a and the corresponding value of technology B is given by b. If the consumers adopt the same technology, they gain the network benefit n.

In the first period, the true stand-alone values are unknown to the users.[51] Therefore, the consumers base their adoption decision on expected values. Assume that the stand-alone values are drawn from a bivariate normal distribution, where $f(a,b)$ represents the joint density function

$$f(a,b) = \frac{1}{2\pi\sigma_A\sigma_B\sqrt{1-\rho^2}}$$
$$\exp\left\{\frac{-1}{2(1-\rho^2)}\left[\left(\frac{a-\mu_A}{\sigma_A}\right)^2 + \left(\frac{b-\mu_B}{\sigma_B}\right)^2 - 2\rho\left(\frac{a-\mu_A}{\sigma_A}\right)\left(\frac{b-\mu_B}{\sigma_B}\right)\right]\right\}. \tag{5.1}$$

and the set of potential stand-alone values corresponds to $\Omega = \{(a,b):(a,b)\in R^2\}$. The mean values of technology A and B are denoted

---

[51] For simplicity, it is assumed that consumers have incomplete information only with respect to the stand-alone values. But the main results also hold if we assume incomplete information about the network benefit and/or the switching costs.

by μ_A and μ_B, and the standard deviations are given by σ_A and σ_B. The correlation coefficient ρ has the following characteristics:

- $-1 < \rho < 1$,
- $\rho = 0$ if A and B are uncorrelated,
- $\rho > 0$ if A and B are positively correlated,
- $\rho < 0$ if A and B are negatively correlated.

The conditional expected values of A and B correspond to

$$E(B|a) = \mu_B + \rho \frac{\sigma_B}{\sigma_A}(a - \mu_A) \text{ and } E(A|b) = \mu_A + \rho \frac{\sigma_A}{\sigma_B}(b - \mu_B) \qquad (5.2)$$

and the joint (cumulative) distribution function is defined as

$$F(a,b) = P(A < a, B < b) = \int_{-\infty}^{a} \int_{-\infty}^{b} f(z,t)dt\,dz, \qquad (5.3)$$

where $F(a,b)$ is non-decreasing in a and b, $F(-\infty,-\infty) = 0$ and $F(\infty,\infty) = 1$.

In the second period, the values of all technologies used in the previous period are revealed. This assumption reflects the idea that potential values of new technologies can be assessed only after the technologies have been used. Based on this information, a user has three options: Firstly, the user can stick to the own technology. As a second option, he may switch to the other technology. Due to technology-specific learning or investment in complementary products switching entails switching costs s. As the third option, the user may forgo to adopt technologies A or B, respectively. The user will choose this "outside option" if both technologies' stand-alone values turn out to be low. By assumption, the user's utility is equal to zero in the case of choosing the outside option.

Fig. 5.1 clarifies the model structure. Experimentation means that the users adopt different technologies in the first period so that they have complete information about the stand-alone values in the second period. The second period of the experimentation subgame involves six distinct Nash-equilibria. If both technologies have similar basic values, each user will stick to his technology and incompatibility will persist. E_{AB} represents all combinations of a and b resulting in this case. In situations where both values turn out to be diverse, the disadvantaged consumer may switch to the superior technology. E_{AA} and E_{BB} stand for the values of a and b yielding ex-post standardization on technology A or B, respectively. E_A and E_B depict the event that the disadvantaged user prefers the outside option while the other consumer sticks to his technology A or B, respectively. Finally, $E_\emptyset$ denotes that a and b turn out to be low so that both users choose the outside option. Thus, the set of events is given by $\Omega = E_{AB} \cup E_{AA} \cup E_{BB} \cup E_A \cup E_B \cup E_\emptyset$ and $\Omega = \{(a,b) : (a,b) \in R^2\}$.

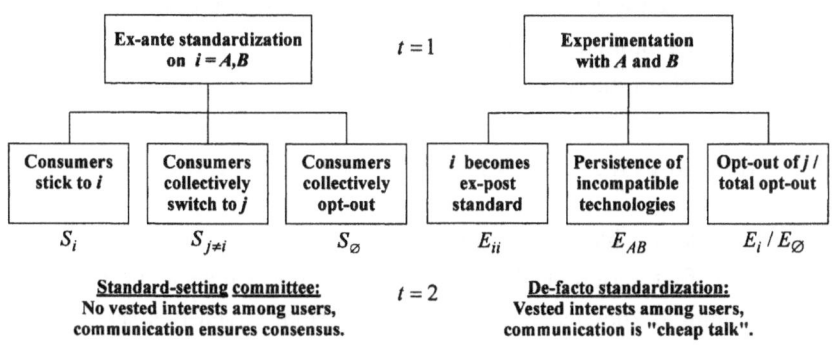

Fig. 5.1: Model structure

As an example for ex-ante standardization, consider the case where both users adopt technology A in the first period. Then, in the second period, they only know the true value of A. But on the basis of the observed value of technology A, the users may derive the conditional expected value of technology B. If this value minus the switching cost is larger than the basic value of A, the users collectively switch to technology B. This event is denoted by S_B. In the case where the users stick to technology A, the equilibrium area is given by S_A. The last event $S_\emptyset$ means that both users choose the outside option, *i.e.* the true value of A and the conditional expected value of B turn out to be low.

If both technologies are equally attractive ex ante, the users have no vested interests in the first period so that they can coordinate their strategies within a committee. However, in the case of an asymmetric distribution of the technologies' values, the first period *may* involve a conflict of interests among users.[52] By adopting different technologies in the first period, consumers build up vested interests, and in the second period of the experimentation subgame, they face a situation corresponding to the Battle of the Sexes. However, in the second period of the ex-ante standardization subgame, users have no vested interests. Thus, there exists a simple coordination problem that can be solved by communication.

5.3   Experimentation

Following the backward-induction principle, we will first derive the equilibria and the socially optimal results of the second period before computing the expected payoffs of the first period.

5.3.1   Second Period: Equilibria of the Experimentation Subgame

Suppose that user 1 has adopted technology A and user 2 has chosen technology B in the first period. Then, the true stand-alone values of both technologies are common knowledge in the second period. Given this scenario, Matrix 5.1 depicts the payoffs of the users depending on the strategies "switch" (S), "no switch" (NS) and "opt out" (OO).

The combination of strategies (NS, S) denotes, for instance, that user 2 switches from technology B to A so that the latter technology becomes ex-post standard. Then, both users realize network benefits, but user 2 has to bear the switching cost alone. The set E_{AA} includes all combinations (a, b) resulting in (NS, S) as Nash-equilibrium. In the case of (NS, OO), user 2 prefers the outside option and realizes the reservation utility equal to zero, while user 1 sticks to technology A. Due to the withdrawal of user 2, user 1 does not obtain network benefits n. Set E_A includes all combinations of stand-alone values inducing (NS, OO) as Nash-equilibrium.

---

[52] The numerical analysis in 5.5.2 will demonstrate that an asymmetric probability distribution does not necessarily involve vested interests.

Matrix 5.1: Payoffs in the experimentation subgame

	Switch (S)	No Switch (NS)	Opt Out (OO)
Switch (S)	$(b-s,\ a-s)$	$(b+n-s,\ b+n)$	$(b-s,\ 0)$
No Switch (NS)	$(a+n,\ a+n-s)$	$(a,\ b)$	$(a,\ 0)$
Opt Out (OO)	$(0,\ a-s)$	$(0,\ b)$	$(0,\ 0)$

It is straightforward to see that the (ex post-) experimentation subgame has the following equilibria for $(a,b) \in R^2$ and $s > n$:

(NS, S):
$$E_{AA} = \{(a,b) : b < n - s + a \wedge a > s - n\}, \tag{5.4a}$$

(S, NS):
$$E_{BB} = \{(a,b) : b > s - n + a \wedge b > s - n\}, \tag{5.4b}$$

(NS, NS):
$$E_{AB} = \left\{ \begin{array}{l} (a,b) : b > n - s + a \ \wedge \\ b < s - n + a \wedge a > 0 \wedge b > 0 \end{array} \right\}, \tag{5.4c}$$

(NS, OO):
$$E_A = \{(a,b) : 0 < a < s - n \wedge b < 0\}, \tag{5.4d}$$

(OO, NS):
$$E_B = \{(a,b) : 0 < b < s - n \wedge a < 0\}, \tag{5.4e}$$

(OO, OO):
$$E_\varnothing = \{(a,b) : a < 0 \wedge b < 0\}. \tag{5.4f}$$

Fig. 5.2 depicts the equilibrium areas for the (a, b)-plane. In order to restrict the analysis to unique Nash-equilibria, switching costs are assumed to exceed the network benefit, *i.e.* $s > n$. For $s < n$, the line $b = s - n + a$ runs below $b = s - n + a$ so that equilibrium regions E_{BB} and E_{AA} overlap.

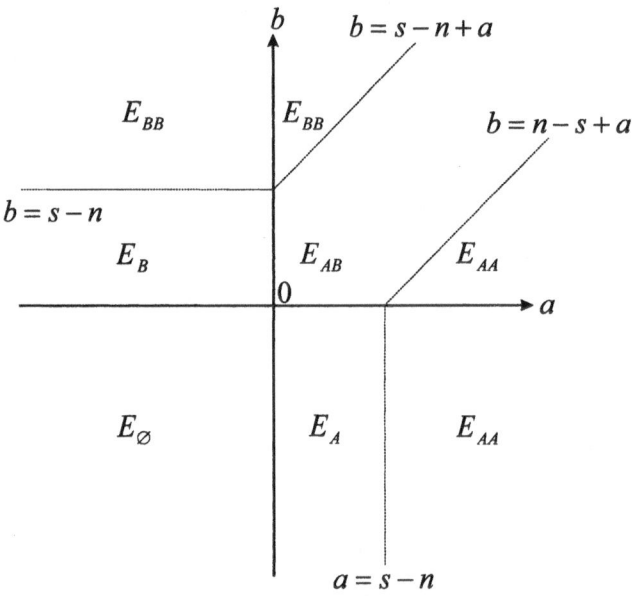

Fig. 5.2: Nash-equilibria for the (a, b)-plane

5.3.2   Second Period: Socially Optimal Outcomes

In this section, we will discuss the socially optimal results of the experimentation subgame. Welfare is derived by means of an additive welfare function which sums up users' payoffs. Welfare analysis will show that the non-cooperative game results in too little ex-post standardization compared with the social optimum. This inefficiency is due to the fact that the switching consumer generates a positive network externality for the other user. Since the switching consumer has to bear the switching cost alone, the private incentive for standardization is too weak compared with the social optimum.

Consider the set W_{AA} which includes all pairs of (a, b) yielding efficient ex-post standardization on A. Comparison of (NS, S) with (NS, NS) results in:

$$a+b<2a+2n-s \implies b<2n-s+a.$$

Comparing (NS, S) with (NS, OO) results in:

$$a<2a+2n-s \implies s-2n<a.$$

Further conditions are not binding in this case.

The other regions for socially optimal outcomes can be derived analogously. It is straightforward to see that the socially optimal rule for $(a,b) \in R^2$ and $s > 2n$ is given by:

(NS, S): $\qquad W_{AA} = \{(a,b) : b < 2n - s + a \wedge a > s - 2n\},$ $\qquad$ (5.5a)

(S, NS): $\qquad W_{BB} = \{(a,b) : b > s - 2n + a \wedge b > s - 2n\},$ $\qquad$ (5.5b)

(NS, NS): $\qquad W_{AB} = \left\{ \begin{matrix} (a,b) : b > 2n - s + a \wedge \\ b < s - 2n + a \wedge a > 0 \wedge b > 0 \end{matrix} \right\},$ $\qquad$ (5.5c)

(NS, OO): $\quad W_A = \{(a,b) : 0 < a < s - 2n \wedge b < 0\},$ $\qquad$ (5.5d)

(OO, NS): $\quad W_B = \{(a,b) : 0 < b < s - 2n \wedge a < 0\},$ $\qquad$ (5.5e)

(OO, OO): $\quad W_{\varnothing} = \{(a,b) : a < 0 \wedge b < 0\}.$ $\qquad$ (5.5f)

In the case of $s < 2n$, the socially optimal rule suggests that the users standardize on the superior technology if and only if this technology exhibits a positive basic value, *i.e.* $a > 0$ or $b > 0$, respectively. Since total network benefits $2n$ exceed switching costs, the equilibria (NS, NS), (NS, OO) and (OO, NS) can never be efficient. The socially optimal areas for $(a,b) \in R^2$ and $s < 2n$ are given by:

$$W_{AA} = \{(a,b) : a > b \wedge a > 0\}, \qquad (5.6a)$$

$$W_{BB} = \{(a,b) : b > a \wedge b > 0\}, \qquad (5.6b)$$

$$W_{AB} = W_A = W_B = \{\ \}, \qquad (5.6c)$$

$$W_{\varnothing} = \{(a,b) : a < 0 \wedge b < 0\}. \qquad (5.6d)$$

Fig. 5.3 shows the socially optimal outcomes for $s > 2n$, which are depicted by: W_{AA}: $d + e + f + g$; W_{BB}: $j + k + l + m$; W_{AB}: h; W_A: c; W_B: p; $W_{\varnothing}$: q. Equilibrium areas are given by: E_{AA}: $e + f$; E_{BB}: $k + l$; E_{AB}: $g + h + j$; E_A: $c + d$; E_B: $m + p$; $E_{\varnothing}$: q.

Comparison of the socially optimal outcomes with the equilibria results in the dotted areas ($d + g$ and $j + m$) which represent the region of inefficiency. In this region, ex-post standardization on A or B (respectively) is socially optimal. In areas j and g, private incentives result in E_{AB} as Nash-equilibrium. In regions d and m, the corresponding equilibria are E_A and E_B, respectively. Thus, private incentives for ex-post standardization are to weak compared with the social optimum.

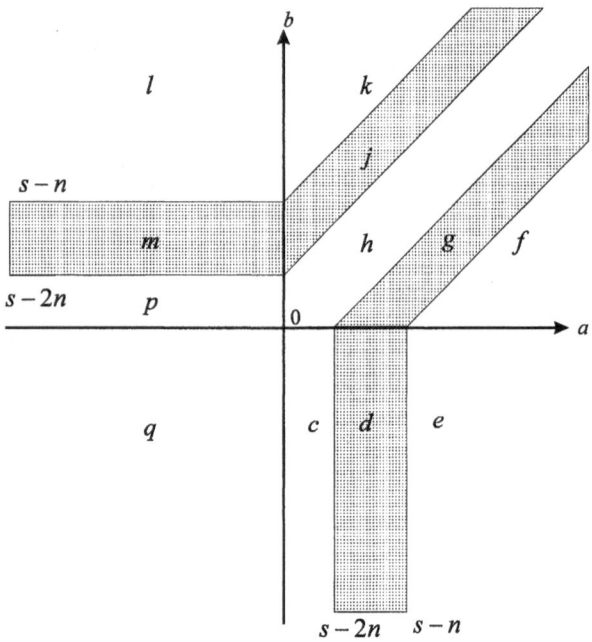

Fig. 5.3: Socially optimal outcomes for $s > 2n$

5.3.3   First Period: Expected Payoffs Without Government Intervention

In the first period, the stand-alone values of both technologies are unknown to the users. Following the backward-induction principle, consumers are assumed to anticipate the potential equilibria of the second period, *i.e.* they weigh the different payoffs with the corresponding probabilities of occurrence. In order to restrict the analysis to a unique equilibrium in $t = 2$, we assume that $s > n$. For the sake of brevity, we will only derive the expected value for experimentation with technology A. The corresponding value for technology B can be computed analogously.

The user who experiments with technology A (the "A-user") realizes the following expected value:

$$V_{E,A} = \mu_A + \delta \int_{E_{AA}} (a+n) f(a,b) \, da \, db$$

$$+ \delta \left[\int_{E_A + E_{AB}} a f(a,b) \, da \, db + \int_{E_{BB}} (b+n-s) f(a,b) \, da \, db \right], \tag{5.7a}$$

$$V_{E,A} = \mu_A + \delta \int_{E_{AA}} n f(a,b) \, da \, db$$

$$+ \delta \left[\int_{E_{AA} + E_A + E_{AB}} a f(a,b) \, da \, db + \int_{E_{BB}} (b+n-s) f(a,b) \, da \, db \right]. \tag{5.7b}$$

The expected value for the first period equals μ_A. Since both consumers adopt different technologies in the first period, they forgo network benefits in $t = 1$. The sum of integrals represents the A-user's expected payoff for the second period. The discount factor is denoted by δ. The first integral in (5.7a) embodies the expected payoff if technology A becomes ex-post standard. The second integral represents the situation where technology A is adopted by the A-user only. Finally, the last integral denotes the expected payoff if the A-user switches to technology B.

Making use of equilibrium conditions (5.4a) – (5.4d), we get

$$V_{E,A} = \mu_A + \delta \int_{s-n}^{\infty} \int_{-\infty}^{n-s+a} n f(a,b) \, db \, da$$

$$+ \delta \left[\int_{0}^{\infty} \int_{-\infty}^{s-n+a} a f(a,b) \, db \, da + \int_{s-n}^{\infty} \int_{-\infty}^{n-s+b} (b+n-s) f(a,b) \, da \, db \right]. \tag{5.8}$$

Regarding the second integral, the integration area $E_{AA} + E_A + E_{AB}$ corresponds to the region below the line $b = s - n + a$ and to the right of the ordinate, as it can be seen in Fig. 5.2. In the case of the last integral, the order of integration is inverted, *i.e.* the integration over E_{BB} is laterally reversed to the integration over E_{AA}.

In the case of a symmetric density function ($\mu = \mu_A = \mu_B$ and $\sigma = \sigma_A = \sigma_B$), we have

$$\int_{E_{BB}} n\, f(a,b)da\, db = \int_{E_{AA}} n\, f(a,b)\, da\, db .$$

Then, both users realize the same expected value of experimentation irrespective of the adopted technology:

$$V_E = \mu + \delta \int_0^\infty \int_{-\infty}^{s-n+a} a\, f(a,b)\, db\, da$$
$$+ \delta \int_{s-n}^\infty \int_{-\infty}^{n-s+b} (b + 2n - s)\, f(a,b)\, da\, db .$$

(5.9)

5.3.4   First Period: Expected Payoffs in the Case of Government Intervention

Suppose that the consumers expect a social planner to intervene in $t = 2$. In this case, the users integrate over the regions of socially optimal outcomes, given by (5.5a) – (5.5d), instead of integrating over the equilibrium areas. For $s > 2n$, the expected value of experimentation is equal to:

$$V_{W,A} = \mu_A + \delta \int_{s-2n}^\infty \int_{-\infty}^{2n-s+a} n\, f(a,b)\, db\, da$$
$$+ \delta \left[\int_0^\infty \int_{-\infty}^{s-2n+a} a\, f(a,b)\, db\, da + \int_{s-2n}^\infty \int_{-\infty}^{2n-s+b} (b + n - s)\, f(a,b)\, da\, db \right].$$

(5.10)

Using a symmetric density function, we get:

$$V_W = \mu + \delta \int_0^\infty \int_{-\infty}^{s-2n+a} a\, f(a,b)\, db\, da$$
$$+ \delta \int_{s-2n}^\infty \int_{-\infty}^{2n-s+b} (b + 2n - s)\, f(a,b)\, da\, db .$$

(5.11)

In the case of $s < 2n$, we make use of (5.6a) and (5.6b):

$$V_{W,A} = \mu_A + \delta \int_0^\infty \int_{-\infty}^a (a+n)f(a,b)\,db\,da$$

$$+ \delta \int_0^\infty \int_{-\infty}^b (b+n-s)f(a,b)\,da\,db. \tag{5.12}$$

It remains to be analyzed whether the intervention of a social planner in $t = 2$ increases the users' expected values in $t = 1$ or not, *i.e.* we will examine whether the ex-post intervention is socially optimal from the ex-ante perspective.

Proposition 5.1: *If the basic values are asymmetrically distributed, the second-period intervention of a social planner may reduce the expected value of a user in $t = 1$, i.e. $V_{E,i} > V_{W,i}$. However, total changes in the expected values are always positive, i.e. $V_{W,A} + V_{W,B} > V_{E,A} + V_{E,B}$.*

Proof: Making use of Fig. 5.3, we can derive the difference $V_{W,A} - V_{E,A}$:

$$V_{W,A} - V_{E,A} = \delta \int_{d+g} n\,f(a,b)\,da\,db$$

$$+ \delta \left[\int_m (b+n-s)\,f(a,b)\,da\,db + \int_j (b-a+n-s)\,f(a,b)\,da\,db \right]. \tag{5.13}$$

Once again, we distinguish between the "A-user" who experiments with technology A and the "B-user" who adopts technology B in the first period. If inefficiency area $d + g$ is given, the social planner forces the B-user to switch to technology A. Thus, the first integral captures the A-user's additional network benefit which results from the enforced switching of the other user.

In the case of inefficiency area m, the A-user has to switch to B instead of choosing the outside option. His change in payoffs is then equal to $b+n-s < 0$, which must be negative in region m. In area j, the A-user is forced to switch to B instead of sticking to A. Then, his change in payoffs equals $b-a+n-s < 0$, which is always negative in region j. Hence, the last two integrals embody the cost of the A-user's enforced switching. In the end, the sign of $V_{W,A} - V_{E,A}$ can be positive or negative.

Analogously, the impact of intervention on the expected value of technology B is equal to:

$$V_{W,B} - V_{E,B} = \delta \int_{m+j} n f(a,b) \, da \, db$$

$$+ \delta \left[\int_d (a+n-s) f(a,b) da \, db + \int_g (a-b+n-s) f(a,b) da \, db \right]. \tag{5.14}$$

We define: $V_{W,t} = V_{W,A} + V_{W,B}$ and $V_{E,t} = V_{E,A} + V_{E,B}$. Summing up $V_{W,A} - V_{E,A}$ and $V_{W,B} - V_{E,B}$ yields:

$$V_{W,t} - V_{E,t} =$$

$$\delta \left[\int_d (a+2n-s) f(a,b) da \, db + \int_g (a-b+2n-s) f(a,b) da \, db \right] \tag{5.15}$$

$$+ \delta \left[\int_m (b+2n-s) f(a,b) da \, db + \int_j (b-a+2n-s) f(a,b) da \, db \right] > 0 .$$

Since all integrals are positive (as it can be easily seen in Fig. 5.3), the total expression is positive as well.   QED

It remains to be shown that the intervention of the social planner in the second period may reduce the expected value of a user in $t = 1$. Fig. 5.4 depicts this case. For $\rho > 0.601379$, $V_{W,A} - V_{E,A}$ is negative, i.e. the user who is experimenting with the more risky technology A is worse off by the intervention in $t = 2$. But as shown in Proposition 5.1, total changes in the expected values are positive.

Proposition 5.2: *If the basic values are symmetrically distributed, the second-period intervention of a social planner increases the expected values of both users in the first period, i.e. $V_W > V_E$.*

Proof: Proposition 5.2 can be directly derived from Proposition 5.1: Using $V_{W,t} > V_{E,t}$, $V_{E,t} = 2V_E$ and $V_{W,t} = 2V_W$, we easily see that $V_W > V_E$ holds in the symmetric case.   QED

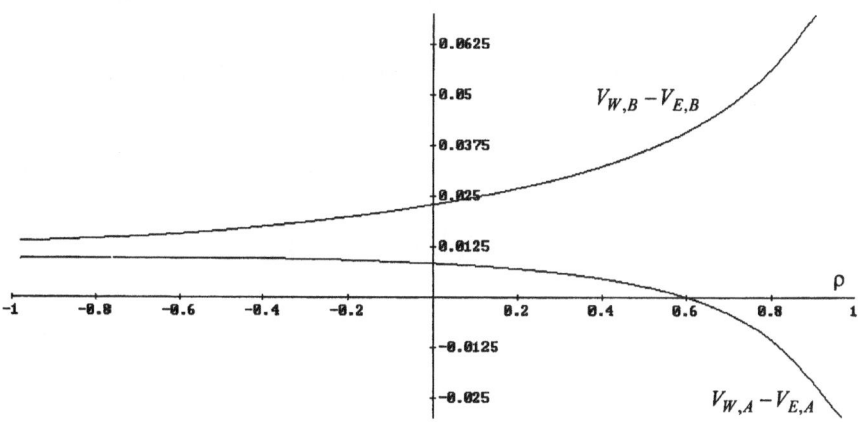

Fig. 5.4: Impact of the ex-post intervention on the expected values for parameters $\mu_A = 1$, $\mu_B = 2$, $\sigma_A = 1.5$, $\sigma_B = 1$, $n = 0.4$, $s = r = 1$

5.4 Ex-Ante Standardization

In the case of ex-ante standardization, users have not built up vested interests in the first period (unlike in the experimentation subgame) and communication is sufficient to ensure consensus in the second period. Following the backward-induction principle, we will first analyze the second-period problem.

5.4.1 Second Period: Equilibria in the Ex-Ante Standardization Subgame

Assume that both users have adopted technology A in the first period, *i.e.* only the true basic value of A is known. But on the basis of the observed value a, the users may derive the conditional expected value of technology B, which is denoted by $E(B|a)$. If the observed value a exceeds $E(B|a) - s$ and $a + n > 0$ holds, both users stick to technology A in the second period and the corresponding equilibrium is given by S_A. On the other hand, both consumers benefit from the collective switch to technology B if $E(B|a)-s$ is larger than the realized value a and $E(B|a)+n-s>0$ holds. This event is denoted by S_B. In situations where both $E(B|a)+n-s$ and $a+n$ turn out to be negative, users prefer the outside option. The corresponding equilibrium is denoted by $S_\varnothing$.

Given the observed value $a \in R$, there are the following equilibria for $t=2$:

$$S_A = \{a : a > E(B|a)-s \wedge a+n > 0\}, \tag{5.16a}$$

$$S_B = \{a : E(B|a)-s > a \wedge E(B|a)+n-s > 0\}, \tag{5.16b}$$

$$S_\varnothing = \{a : a+n < 0 \wedge E(B|a)+n-s < 0\}. \tag{5.16c}$$

Depending on the slope and location of $E(B|a)+n-s$, six cases must be distinguished (see Matrix 5.2). We define ε as location parameter which shows (together with ρ) whether $E(B|a)+n-s>0$ at $a=-n$ or not.

$$E(B|(a=-n))+n-s = \mu_B - \rho\frac{\sigma_B}{\sigma_A}(n+\mu_A)+n-s > 0,$$

$$\Rightarrow \quad \rho < \varepsilon = \frac{\sigma_A}{\sigma_B}\frac{\mu_B+n-s}{\mu_A+n}. \tag{5.17}$$

Analogously, $\rho > \varepsilon$ implies that $E(B|a)+n-s < 0$ at $a = -n$.

We define a_0 as the point where the line $E(B|a)+n-s$ intersects the line $a+n$:

$$E(B|a_0)+n-s = a+n,$$

$$\Rightarrow \quad a_0 = \frac{\rho\sigma_B\mu_A - \sigma_A(\mu_B-s)}{\rho\sigma_B - \sigma_A}, \quad \text{not defined for } \rho = \frac{\sigma_A}{\sigma_B}. \tag{5.18}$$

Moreover, we define a_1 as the point where the line $E(B|a)+n-s$ intersects the abscissa:

$$E(B|a_1)+n-s = 0,$$

$$\Rightarrow \quad a_1 = \frac{\rho\sigma_B\mu_A - \sigma_A(\mu_B+n-s)}{\rho\sigma_B}, \quad \text{not defined for } \rho = 0. \tag{5.19}$$

Matrix 5.2: Distinction of cases for the ex-ante standardization subgame in t = 2

	Location of the line $E(B\|a)+n-s$	
Slope of the line $E(B\|a)+n-s$	Positive value at $a = -n$ $\rho < \varepsilon$	Negative value at $a = -n$ $\rho > \varepsilon$
negative $-1 < \rho < 0$	Case 1	Case 2
positive / smaller than 1 $0 < \rho < \sigma_A / \sigma_B$	Case 3	Case 4
positive / larger than 1 $\sigma_A / \sigma_B < \rho < 1$	Case 5	Case 6

Case 1: $-1 < \rho < 0$ and $\rho < \varepsilon$

Fig. 5.5 depicts the second-period equilibria for the situation where both users have adopted technology A before. The line $E(B|a)+n-s$ has a negative slope and $E(B|a)+n-s > 0$ holds at $a = -n$.

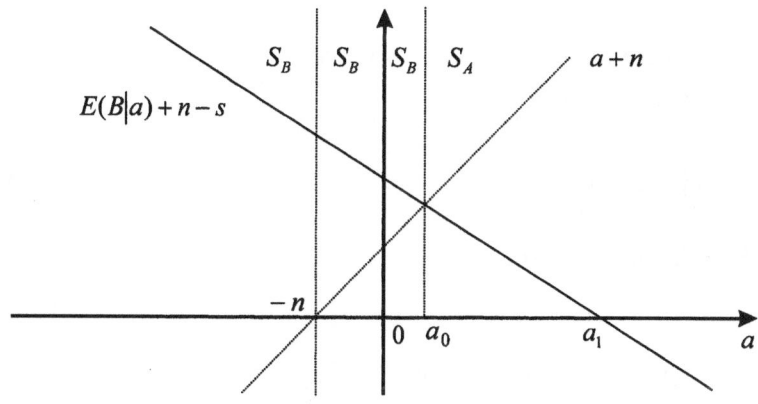

Fig. 5.5: Second-period equilibria for case 1

There are the following equilibria in $t = 2$:

$$S_A = \{a : a > a_0\},$$ (5.20a)

$$S_B = \{a : a < a_0\}.$$ (5.20b)

Case 2: $-1 < \rho < 0$ and $\rho > \varepsilon$

In the second case, the line $E(B|a) + n - s$ has a negative slope and $E(B|a) + n - s < 0$ holds at $a = -n$. Fig. 5.6 depicts this case.

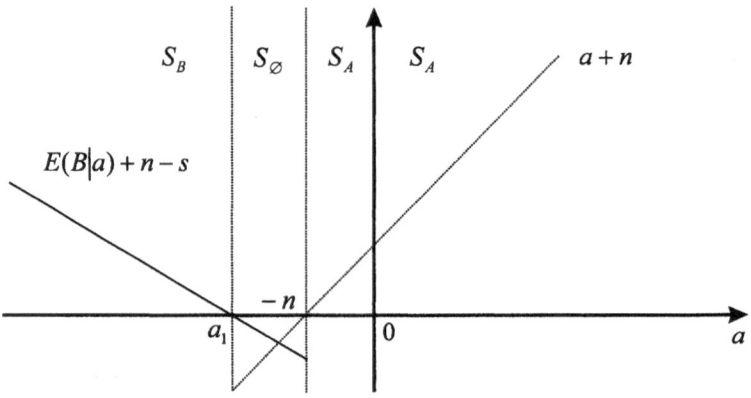

Fig. 5.6: Second-period equilibria for case 2

In $t = 2$, there are the following equilibria:

$$S_A = \{a : a > -n\},$$ (5.21a)

$$S_B = \{a : a < a_1\},$$ (5.21b)

$$S_\emptyset = \{a : a_1 < a < -n\}.$$ (5.21c)

Case 3: $0 < \rho < \sigma_A / \sigma_B$ and $\rho < \varepsilon$

Fig. 5.7 illustrates case 3. The line $E(B|a) + n - s$ has a positive slope which is smaller than one. Moreover, $E(B|a) + n - s > 0$ holds at $a = -n$.

There are the following equilibria in $t = 2$:

$$S_A = \{a : a > a_0\},$$ (5.22a)

$$S_B = \{a : a_1 < a < a_0\},$$ (5.22b)

$$S_\emptyset = \{a : a < a_1\}.$$ (5.22c)

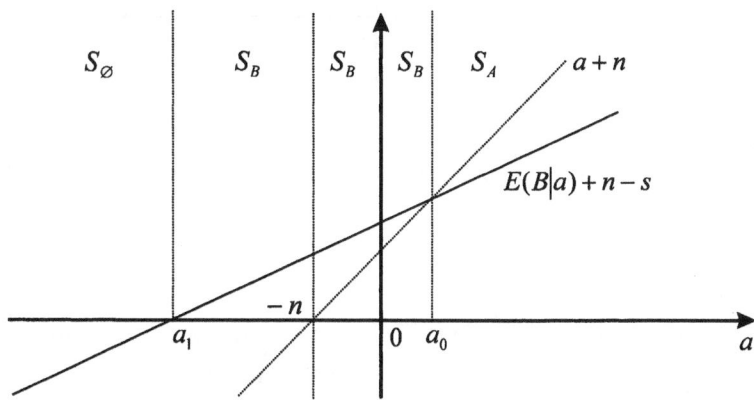

Fig. 5.7: Second-period equilibria for case 3

Case 4: $0 < \rho < \sigma_A / \sigma_B$ and $\rho > \varepsilon$

In this case, the line $E(B|a) + n - s$ has a positive slope which is smaller than one. Furthermore, $E(B|a) + n - s < 0$ holds at $a = -n$. Fig. 5.8 depicts this situation.

In $t = 2$, we have the following equilibria:

$$S_A = \{a : a > -n\},$$ (5.23a)

$$S_\emptyset = \{a : a < -n\}.$$ (5.23b)

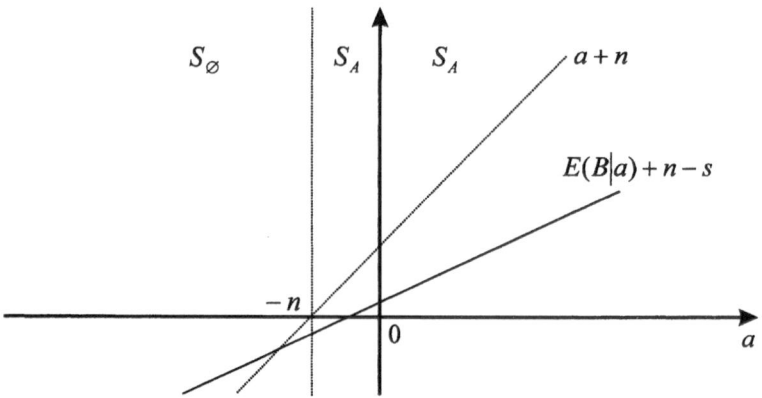

Fig. 5.8: Second-period equilibria for case 4

Case 5: $\sigma_A / \sigma_B < \rho < 1$ and $\rho < \varepsilon$

Fig. 5.9 depicts case 5. The line $E(B|a) + n - s$ has a positive slope which exceeds one. Furthermore, $E(B|a) + n - s > 0$ holds at $a = -n$.

There are the following equilibria in $t = 2$:

$$S_B = \{a : a > a_1\}, \tag{5.24a}$$

$$S_\emptyset = \{a : a < a_1\}. \tag{5.24b}$$

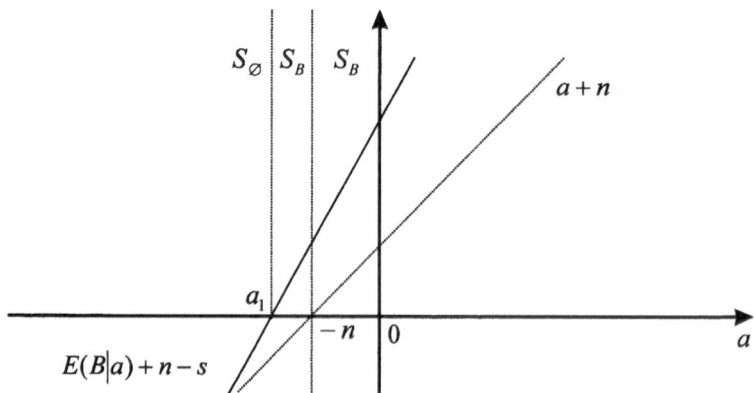

Fig. 5.9: Second-period equilibria for case 5

Case 6: $\sigma_A / \sigma_B < \rho < 1$ and $\rho > \varepsilon$

The last case is depicted in Fig. 5.10. The $E(B|a) + n - s$-line has a positive slope which exceeds one. Moreover, $E(B|a) + n - s < 0$ holds at $a = -n$.

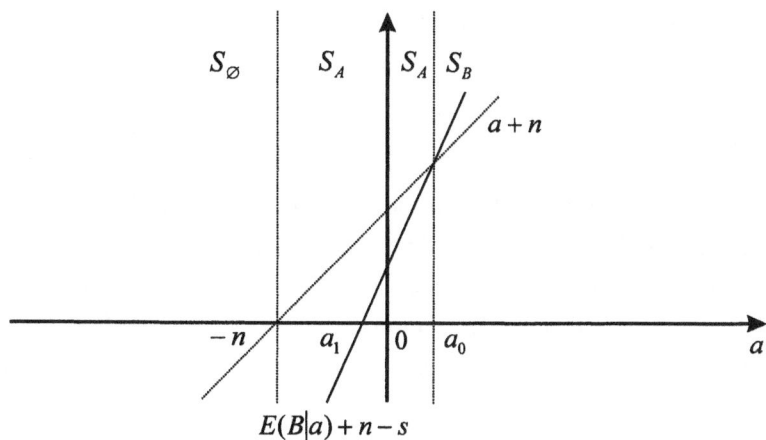

Fig. 5.10: Second-period equilibria for case 6

There are the following equilibria for $t = 2$:

$$S_A = \{a : -n < a < a_0\}, \tag{5.25a}$$

$$S_B = \{a : a > a_0\}, \tag{5.25b}$$

$$S_\emptyset = \{a : a < -n\}. \tag{5.25c}$$

In the case of a symmetric density function, the slope of the line $E(B|a) + n - s$ can never exceed one. Thus, cases 5 and 6 will not apply.

5.4.2   First Period: Expected Payoffs in the Case of Ex-Ante Standardization

Suppose that the users collectively adopt technology A in the first period. At the time of adoption, the basic values of both technologies are still unknown. Following the backward-induction principle, the users anticipate the equilibria of the second period, *i.e.* they weigh the different events such as S_A, S_B and

$S_\varnothing$ with the corresponding probability of occurrence. The expected value of ex-ante standardization on B can be derived completely analogously.

The expected value of standardization on technology A (in its most abstract form) is equal to:

$$V_{S,A} = \mu_A + n + \delta \int_{S_A} (a+n) f(a,b) \, db \, da$$
$$+ \delta \int_{S_B} (b+n-s) f(a,b) \, db \, da.$$

(5.26)

The expected stand-alone value for the first period is equal to μ_A. Due to ex-ante standardization, the users realize network benefits from the beginning. The expression in the square brackets represents the expected payoff for the second period. The first integral embodies the expected payoff if the users stick to technology A. The second integral stands for the case where the users collectively switch to technology B. Since the reservation utility is normalized to zero, the integration over $S_\varnothing$ is omitted.

Depending on the parameters of the normal distribution, we make use of the six cases. Substitution of conditions (5.20) – (5.25) yields:

Case 1: $-1 < \rho \le 0 \ \wedge \ \rho \le \varepsilon$

$$V_{S,A} = \mu_A + n + \delta \int_{a_0}^{\infty} \int_{-\infty}^{\infty} (a+n) f(a,b) \, db \, da$$
$$+ \delta \int_{-\infty}^{a_0} \int_{-\infty}^{\infty} (b+n-s) f(a,b) \, db \, da.$$

(5.27a)

Case 2: $-1 < \rho < 0$ and $\rho \ge \varepsilon$

$$V_{S,A} = \mu_A + n + \delta \int_{-n}^{\infty} \int_{-\infty}^{\infty} (a+n) f(a,b) \, db \, da$$
$$+ \delta \int_{-\infty}^{a_1} \int_{-\infty}^{\infty} (b+n-s) f(a,b) \, db \, da.$$

(5.27b)

Case 3: $0 < \rho < \sigma_A / \sigma_B$ and $\rho \le \varepsilon$

$$V_{S,A} = \mu_A + n + \delta \int_{a_0}^{\infty} \int_{-\infty}^{\infty} (a+n) f(a,b) \, db \, da$$

$$+ \delta \int_{a_1}^{a_0} \int_{-\infty}^{\infty} (b+n-s) f(a,b) \, db \, da \ . \tag{5.27c}$$

Case 4: $0 \le \rho \le \sigma_A / \sigma_B$ and $\rho \ge \varepsilon$

$$V_{S,A} = \mu_A + n + \delta \int_{-n}^{\infty} \int_{-\infty}^{\infty} (a+n) f(a,b) \, db \, da \ . \tag{5.27d}$$

Case 5: $\sigma_A / \sigma_B \le \rho < 1$ and $\rho \le \varepsilon$

$$V_{S,A} = \mu_A + n + \delta \int_{a_1}^{\infty} \int_{-\infty}^{\infty} (b+n-s) f(a,b) \, db \, da \ . \tag{5.27e}$$

Case 6: $\sigma_A / \sigma_B < \rho < 1$ and $\rho \ge \varepsilon$

$$V_{S,A} = \mu_A + n + \delta \int_{-n}^{a_0} \int_{-\infty}^{\infty} (a+n) f(a,b) \, db \, da$$

$$+ \delta \int_{a_0}^{\infty} \int_{-\infty}^{\infty} (b+n-s) f(a,b) \, db \, da \ . \tag{5.27f}$$

5.5   Ex-Ante Standardization vs. Experimentation

Suppose that the expected values μ_A and μ_B are positive so that it is never worthwhile to opt out in $t = 1$. The payoffs for user 1 and 2 are depicted in Matrix 5.3. It is straightforward to see that the first-period game has the following Nash-equilibria:

$(A, \ A)*$ $\qquad\qquad$ if $V_{E,B} - V_{S,A} < 0$, $\qquad\qquad$ (5.28a)

$(B,\ B)*$ if $V_{E,A} - V_{S,B} < 0$,   (5.28b)

$(A,\ B)*$ and $(B,\ A)*$ if $V_{E,A} - V_{S,B} > 0$ and $V_{E,B} - V_{S,A} > 0$.   (5.28c)

Note that experimentation always involves multiplicity. However, in the case of ex-ante standardization on A or B, a unique equilibrium may exist. For instance, the equilibrium $(A,\ A)*$ is unique if both $V_{E,B} - V_{S,A} < 0$ and $V_{E,A} - V_{S,B} > 0$ hold.

For a symmetric density function, we have $V_E = V_{E,A} = V_{E,B}$ and $V_S = V_{S,A} = V_{S,B}$. Thus, there is no clash of interests in the first period and communication is sufficient to ensure uniqueness.

Matrix 5.3: Expected payoffs in $t = 1$

	Technology A	Technology B
Technology A	$(V_{S,A}\ ,\ V_{S,A})$	$(V_{E,A}\ ,\ V_{E,B})$
Technology B	$(V_{E,B}\ ,\ V_{E,A})$	$(V_{S,B}\ ,\ V_{S,B})$

Suppose that a social planner may intervene in both periods. In $t = 2$, the social planner prevents the ex-post inefficiency in the experimentation sub-game. As already shown in Proposition 5.1, this ex-post intervention (if anticipated by the users) increases the sum of expected values. But the intervention in $t = 2$ is not sufficient to ensure socially optimal outcomes, i.e. the social planner should also intervene in the first period in order to prevent inefficient ex-ante decisions.

The socially optimal first-period outcomes are given by:

$(A,\ A)$ if $2V_{S,A} > V_{W,A} + V_{W,B}$ $\wedge$ $V_{S,A} > V_{S,B}$,   (5.29a)

$(B,\ B)$ if $2V_{S,B} > V_{W,A} + V_{W,B}$ $\wedge$ $V_{S,B} > V_{S,A}$,   (5.29b)

$(A,\ B)$ if $V_{W,A} + V_{W,B} > 2V_{S,A}$ $\wedge$ $V_{W,A} + V_{W,B} > 2V_{S,B}$.   (5.29c)

5.5.1   Numerical Examples for a Symmetric Density Function

In this section, we will analyze how parameters such as s, n, μ, σ and ρ affect the users' first-period decisions.

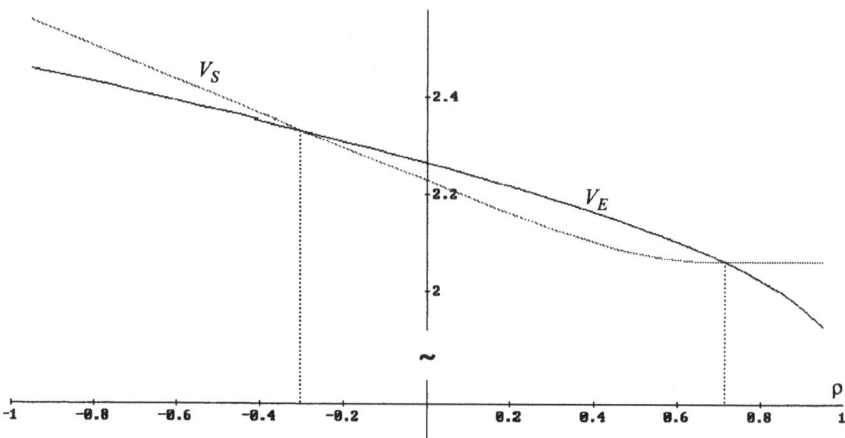

Fig. 5.11: Ex-ante standardization vs. experimentation for parameters $\mu = \sigma = 0.9$, $s = 0.3$, $n = 0.1$, $\delta = 1$

Fig. 5.11 compares the expected values of experimentation and ex-ante standardization.[53] Both values decline with increasing correlation. If the values are strongly correlated, ex-ante standardization is superior to experimentation. An intuitive explanation for this result is due to different information effects of both strategies. In the case of ex-ante standardization, the switching deci-sion in $t = 2$ is based on limited information because the users know the true value of the joint technology only. But if the values are strongly correlated, the consumers may use the observed value in order to learn something about the other value, $i.e.$ they revise the expected ex-ante value of the other technology according to the Bayesian rule. Thus, experimentation only has a slight infor-mation advantage over ex-ante standardization in the case of strong correla-tion. However, in the case of uncorrelated technologies, $\rho = 0$, the line $E(B|a) + n - s$ is parallel to the abscissa, $i.e.$ the expected value for B is con-

---

[53] The computation of V_S is based upon the following cases: Case 1 for $-1 < \rho \leq 0$, case 3 for $0 < \rho \leq 0.7$ and case 4 for $0.7 \leq \rho < 1$.

stant irrespective of the realized value of A. Note that for $0.7 \leq \rho < 1$, case 4 applies. Then, switching is not worthwhile because the users expect the values of both technologies to be very similar. Thus, correlation has no impact on the expected value of ex-ante standardization, *i.e.* V_S remains constant.

Fig. 5.12 depicts the impact of increased network effects on $(V_E - V_S)$.[54] If n rises, the comparative advantage of experimentation is reduced. The reason is that n has a stronger impact on V_S because in the case of ex-ante standardization, the users already realize network benefits in the first period.

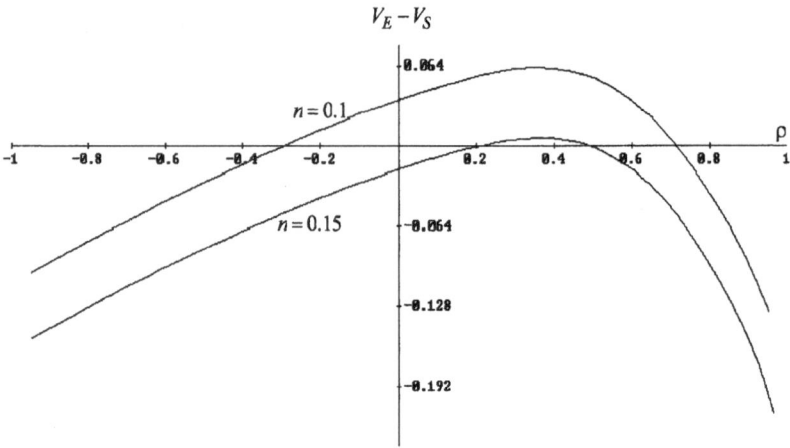

Fig. 5.12: Impact of network effects on $(V_E - V_S)$ for parameters $\mu = \sigma = 0.9$, $s = 0.3$, $\delta = 1$

Fig. 5.13 illustrates the impact of increased switching costs on $(V_E - V_S)$. With increasing switching costs the comparative advantage of experimentation is reduced. An intuitive explanation is that the switching option is more useful to the consumers in the case of experimentation because switching is based on complete information. If switching becomes more expensive, this must have a stronger negative impact on V_E than on V_S.[55]

---

[54] The computation of V_S ($n = 0.15$) involves the following cases: Case 1 for $-1 < \rho \leq 0$, case 3 for $0 < \rho \leq 5/7$ and case 4 for $5/7 \leq \rho < 1$.

[55] The computation of V_S ($s = 0.4$) involves the following cases: Case 1 for $-1 < \rho \leq 0$, case 3 for $0 < \rho \leq 0.6$ and case 4 for $0.6 \leq \rho < 1$.

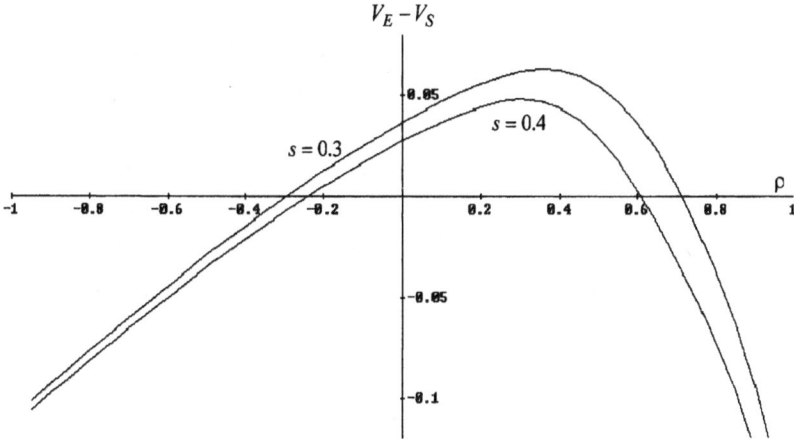

Fig. 5.13: Impact of switching costs on $(V_E - V_S)$ for parameters $\mu = \sigma = 0.9$, $n = 0.1$, $\delta = 1$

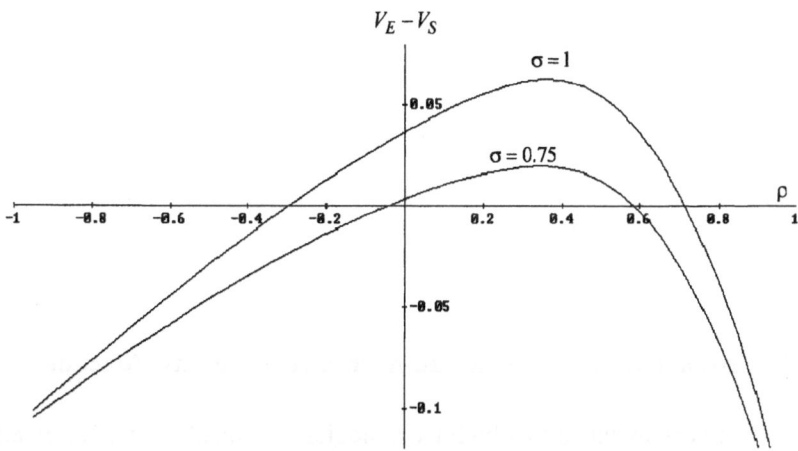

Fig. 5.14: Impact of standard deviation on $(V_E - V_S)$ for parameters $\mu = 0.9$, $s = 0.3$, $n = 0.1$, $\delta = 1$

The impact of standard deviation on $(V_E - V_S)$ is depicted in Fig. 5.14. If σ is lowered, the comparative advantage of experimentation is reduced as well. An intuitive explanation for this result is that larger deviation makes both the switching option and the outside option more important. Both options are more useful in the case of experimentation due to complete information. Note that the difference between the $(V_E - V_S)$ curves is maximal in the case of low positive correlation.[56]

Fig. 5.15 illustrates the impact of reduced mean values on $(V_E - V_S)$. For negative and low positive correlation, the comparative advantage of experimentation becomes stronger. The maximum of the $(V_E - V_S)$ curve shifts to the left, *i.e.* experimentation becomes more attractive for lower values of ρ.[57]

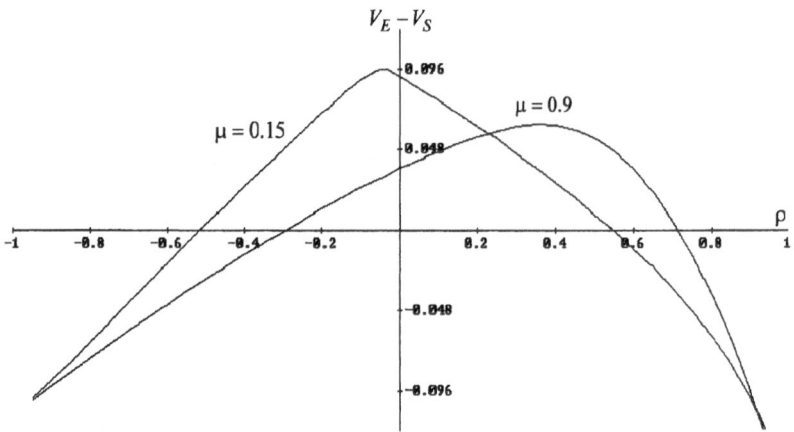

Fig. 5.15: Impact of mean values on $(V_E - V_S)$ for parameters $\sigma = 0.9$, $s = 0.3$, $n = 0.1$ and $\delta = 1$

5.5.2   Numerical Example for an Asymmetric Density Function

So far, we have assumed that both technologies are equally attractive ex ante. In this section, we make the assumption that technology *B* has a larger standard deviation but a lower mean value than *A*, *i.e.* *B* is more risky. Due to the low mean value, users could be reluctant to experiment with *A* even though

---

[56] The computation of V_S ($\sigma = 0.75$) involves the following cases: Case 1 for $-1 < \rho \leq 0$, case 3 for $0 < \rho \leq 0.7$ and case 4 for $0.7 \leq \rho < 1$.

[57] The computation of V_S ($\mu = 0.15$) involves the following cases: Case 1 for $-1 < \rho \leq -0.2$, case 2 for $-0.2 \leq \rho < 0$ and case 4 for $0 \leq \rho < 1$.

experimentation might be worthwhile collectively. The computation of so-cially optimal results is based on the assumption that the social planner may intervene both in the ex-post and ex-ante standardization process. Suppose that users anticipate the social planner's intervention in the ex-post standardization process, *i.e.* the expected values of experimentation are given by $V_{W,i}$ instead of $V_{E,i}$.

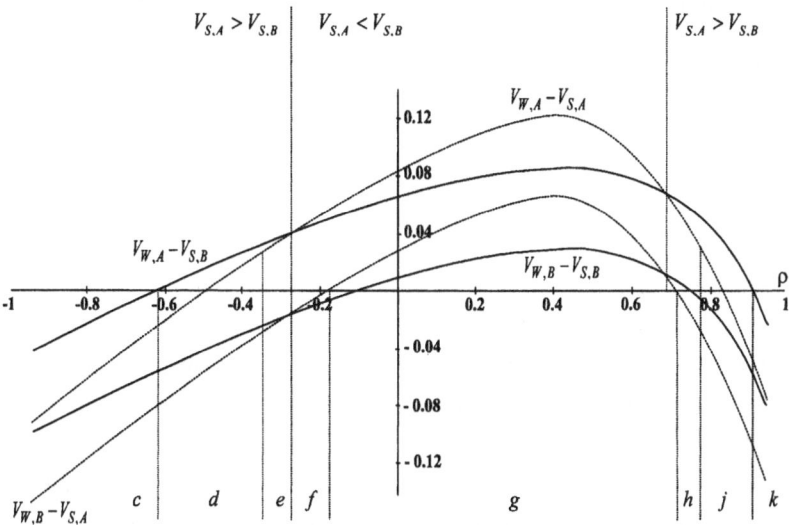

Fig. 5.16: Equilibria and welfare regions for parameters $\mu_A = 1.05$, $\mu_B = 1$, $\sigma_A = 0.8$, $\sigma_B = 1$, $s = 0.2$, $n = 0.1$, $\delta = 1$

Fig. 5.16 depicts the equilibrium regions for a numerical example assuming that technology B has a larger standard deviation but a lower mean value than A. [58] Given that users anticipate the social planner's intervention in the second period, first-period equilibria are as follows:

- In regions *c* and *k*, both $(A, A)^*$ and $(B, B)^*$ exist because $V_{W,A} - V_{S,B} < 0$ and $V_{W,B} - V_{S,A} < 0$ are given.

---

[58] The computation of $V_{S,A}$ involves the following cases: Case 1 for $-1 < \rho \le 0$, case 3 for $0 < \rho \le 72/115$, Case 4 for $72/115 \le \rho \le 4/5$ and case 6 for $4/5 < \rho < 1$. For $V_{S,B}$ there are two cases: Case 1 for $-1 < \rho \le 0$ and case 3 for $0 < \rho < 1$.

- In regions $d + e + f$ and $h + j$, the equilibrium $(A, \ A)^*$ is unique because $V_{W,A} - V_{S,B} > 0$ and $V_{W,B} - V_{S,A} < 0$ hold.

- In region g, $V_{W,A} - V_{S,B} > 0$ and $V_{W,B} - V_{S,A} > 0$ are given so that $(A, \ B)^*$ and $(B, \ A)^*$ exist.

Making use of (5.29), we can derive the socially optimal outcomes in the ex-ante standardization process:

- In regions $c + d$ and $j + k$, ex-ante standardization on A, i.e. $(A, \ A)^*$, is socially optimal because $2V_{S,A} > V_{W,A} + V_{W,B}$ and $V_{S,A} > V_{S,B}$ hold.

- In area $e + f + g + h$, experimentation is socially desirable due to $2V_{S,A} < V_{W,A} + V_{W,B}$ and $2V_{S,B} < V_{W,A} + V_{W,B}$.

Comparing the Nash-equilibria with the socially optimal results, we can identify two sources of inefficiency: Firstly, a coordination failure may occur, i.e. consumers standardize on the wrong technology ex ante. In regions c and k, $(B, \ B)^*$ constitutes an equilibrium even though ex-ante standardization on A would make both consumers better off. Note that $V_{S,A} > V_{S,B} > V_{W,A} > V_{W,B}$ applies in this situation, i.e. both users would benefit from switching collectively to A but each consumer would be worse off by switching, given that the other user sticks to technology B. Since the consumers have no vested interests in this situation, communication could exclude this inefficient result.

The second source of inefficiency is excessive ex-ante standardization on A. In regions $e + f$ and h, experimentation is socially optimal but due to the low mean of technology B, the single consumer is reluctant to experiment with B. Note that the B-user generates a positive information externality for the consumer of A, but this effect is ignored in the private decision problem. A peculiarity occurs in region f. There is not only excessive ex-ante standardization compared with experimentation but also ex-ante standardization on the inferior technology A because $V_{S,B} > V_{S,A}$ holds in this case.

5.6 Conclusions

Some economists hold that early standardization is likely to give the privately and socially best results (Farrell and Saloner, 1988, p. 239). However, we have shown that quality may suffer for speed's sake. Abstaining from early (ex-ante) standardization, users may experiment with diverse technologies to learn about the true values. As a consequence, the ex-post standardization process is based on better information. On the other hand, experimentation involves a transient or even persistent loss of compatibility.

The numerical analysis of this tradeoff is based on the assumption that the values of both technologies are drawn from a bivariate normal distribution. This type of probability distribution has enabled us to study the impact of correlation, as well as the impact of different standard deviations and mean values on the standardization process. We have shown that the users prefer ex-ante standardization to experimentation if they expect the values of both technologies to be strongly correlated. This result can be traced back to the fact that the consumers may use the observed value of the joint technology to learn something about the other value. Then, the information advantage of experimentation is easily outweighed by the compatibility advantage of ex-ante standardization.

The welfare analysis has shown that optimal decisions should be based on the information available at the time of decision-making rather than on ex-post judgements. Consequently, the model distinguishes between the optimal decision ex-ante and the optimal decision ex-post. We have shown that there is too little ex-post standardization in the experimentation subgame due to the fact that consumers have built up vested interests in the first period. The intervention of a social planner in the second period *can* reduce the expected payoff of a user in the first period, but it always increases the sum of expected payoffs, as shown in Proposition 5.1. If technologies are not equally attractive ex ante (*i.e.* there is an asymmetric probability distribution), the technology-adoption process of the first period may be inefficient as well. There can be too much ex-ante standardization compared with the social optimum or, alternatively, the consumers may adopt the inferior technology as standard.

6

Summary of Findings

Three essays on various aspects of standardization and expectations have been presented. The first essay has given economic reasons why university examinations should be standardized. In a first step, we have derived the result that signaling may serve as a job-matching device, thereby allocating heterogeneous employees to the adequate firm type. In contrast to the basic signaling model by Spence (1973), our job-matching framework has shown that signaling may increase total output. However, the approach has suggested that the more productive employees (type B) have too strong incentives for signaling. This inefficiency can be traced back to the fact that type B employees not only take into account the positive job-matching effect of signaling, they also internalize the welfare-neutral distributive effect of signaling. By means of signaling, they distinguish themselves from the less productive type A so that they avoid "subsidizing" type A, as it would be the case with a uniform wage equal to the average productivity of all employees. The equilibrium analysis has shown that relatively low signaling costs result in a unique signaling-separating equilibrium (*SSE*). However, relatively high signaling costs involve multiple equilibria, *i.e.* the SSE, the non-signaling pooling equilibrium (*NSPE*), and an equilibrium in mixed strategies exist. Since the *NSPE* would make all type B employees better off, the *SSE* can be interpreted as a coordination failure in this situation. For a modified approach with a small number of employees (two type A and two type B employees), we have found that signaling becomes less important as distributive device because type B employees can "signal" their higher productivity by means of their impact on total output.

Building on the job-matching approach, the second part of the first essay has dealt with standardization of examination requirements. We have referred to the first main question of how standards may affect employers' and employees' expectations. In our framework, standardization is considered as a means of reducing variation in examination requirements among universities, for example by means of central examinations or accreditation. The numerical analysis has shown that a unique Bayesian equilibrium exists for different levels of deviations in requirements. As demonstrated in the welfare analysis, standardization of requirements improves the job-matching function of education, thereby increasing the expected total output. However, there exists a tradeoff between the job-matching function and total signaling costs. We have found that full standardization of requirements is desirable for low signaling costs. If signaling costs are relatively high, society is better off with maximal deviation in requirements.

The second essay has studied the competition between two firms when their incompatible technologies exhibit network effects. The essay has distinguished two different regimes of standardization. Intra-technology competition involves that firms compete within a joint network (*i.e.* standard), whereas inter-technology competition refers to de-facto standardization by means of deterred or blockaded entry of the rival technology. The essay has put the emphasis on the question of whether an incumbent firm has an incentive to keep its technology for itself, which would result in inter-technology competition, or to share its technology with the rival firm. The second option of sponsoring intra-technology competition, enables the incumbent firm to commit credibly to a future network size which exceeds the profit-maximizing monopoly quantity. This credible commitment increases consumers' willingness to pay for the network good.

The first part of the second essay has dealt with inter-technology competition. The analysis has suggested that the fulfilled expectations equilibria hinge on the relative marginal costs of both technologies. In the case of strong cost differences, the weaker firm's entry is blockaded. For weak cost differences, both firms coexist in an incompatible and heterogeneous duopoly. If the incumbent firm has a moderate cost advantage, it may deter the rival technology's entry. The central argument of the essay is that the deterrence quantity not only prevents the rival's entry, it also serves as commitment to a larger network (compared to the monopoly case with blockaded entry), thereby increasing consumers' willingness to pay for the network good. We have found that the deterrence profit may exhibit perverse cost effects, *i.e.* it may rise with decreasing marginal costs of the rival. This is due to the fact that the incumbent firm has to supply a larger deterrence quantity with increasing strength of its competitor so that it can credibly commit to a larger network and thus spurring consumers' willingness to pay for the network good.

The second part of the essay has been devoted to the incumbent's choice of whether to share its technology with the rival or to insist on inter-technology competition. We have shown that the deterrence strategy can be more profitable for the incumbent if the incumbent's cost advantage is moderate. Since the incumbent does not completely internalize positive externalities of sponsoring intra-technology competition, which exist in terms of the follower profit and in an increased consumers' surplus, entry deterrence is welfare-inferior.

The third essay has dealt with standardization of nascent technologies. In a two-period framework with two competing network technologies and two consumers, we have shown that there exists a tradeoff between early (ex-ante) standardization and experimentation. Experimentation enables users to learn about the stand-alone values of both technologies after using them ("learning by using"). As a consequence the ex-post standardization process is based on better information. While experimentation improves consumers' information, it involves a loss of compatibility in the first period, which may persist in the second period due to vested interests. By means of early (ex-ante) standardization, consumers enjoy network benefits from the beginning, but they forgo information about the alternative technology. Thus, a major conclusion is that early standardization does not necessarily give privately and socially best results, *i.e.* the adoption's quality may suffer for speed's sake.

The third essay has also been devoted to the problem of how consumers' expectations affect the evolution of standards. In order to get traceable results, we have assumed that the technologies' values are drawn from a bivariate normal distribution. As shown in the numerical analysis, consumers prefer ex-ante standardization to experimentation if they expect the values of both technologies to be strongly correlated. This result is due to the fact that consumers use the observed value of the joint technology to learn something about the other value, *i.e.* they revise the expected ex-ante value of the alternative technology according to the Bayesian rule.

Our welfare analysis has distinguished between optimal ex-ante and ex-post decisions. We have demonstrated that there is too little ex-post standardization in the experimentation subgame due to vested interests among consumers. Moreover, we have shown that the first-period adoption decision can be inefficient if technologies are not equally attractive ex-ante. There can be too much ex-ante standardization compared with the social optimum, or consumers may adopt the inefficient technology as ex-ante standard.

A 1

Appendix to Chapter 3

Table A 1.1: Approximated equilibrium values for case 1

d	$P(S|A)*$	$P(S|B)*$	$P(S)*$	$\lambda*$	$\psi*$	y_A*
0.5	0.25	0.75	0.5	0.75	0.25	0.75
0.6	0.231183	0.607528	0.419355	0.724359	0.337963	0.677420
0.7	0.229074	0.524647	0.376860	0.696076	0.381418	0.620703
0.8	0.236509	0.477514	0.357012	0.668766	0.406295	0.578414
0.9	0.248570	0.451321	0.349946	0.644844	0.422025	0.547427
1	0.262621	0.437701	0.350161	0.625	0.432645	0.525241

Table A 1.2: Approximated equilibrium values for case 2

d	$P(S|A)^*$	$P(S|B)^*$	$P(S)^*$	λ^*	ψ^*	y_A^*
0	0	1	0.5	1	0	1
0.1	0.048462	1	0.524231	0.953778	0	0.909692
0.2	0.092931	1	0.546465	0.914971	0	0.837172
0.2996	0.132450	1	0.566225	0.883041	0	0.779762
0.3	0.132844	0.999184	0.566014	0.882650	0.000940	0.779706
0.4	0.206346	0.843911	0.525129	0.803528	0.164349	0.765077
0.5	0.25	0.75	0.5	0.75	0.25	0.75

Table A 1.3: Welfare implications of standardization

d	$C(d)$	$X(d)$	$W(d)$
0	0.3	1.5	1.2
0.1	0.32193	1.47689	1.15496
0.2	0.33804	1.45749	1.11945
0.2996	0.34901	1.44152	1.09251
0.3	0.34882	1.44133	1.09250
0.4	0.30779	1.40176	1.09398
0.5	0.275	1.375	1.1
0.6	0.20161	1.33102	1.12941
0.7	0.15775	1.30929	1.15154
0.8	0.12940	1.29685	1.16745
0.9	0.10877	1.28899	1.18022
1	0.09196	1.28368	1.19172

Appendix to Chapter 4:
Proof of Proposition 8

First, we will demonstrate that $CS^e_{A,C}* > CS^e_{A,D}*(c_{B,2})$ holds:

$$CS^e_{A,C}* > CS^e_{A,D}*(c_{B,2}) \quad \Rightarrow \quad \frac{9(1-n)(\alpha_A - c_A)^2}{2(3n-4)^2} > \frac{2(1-n)(\alpha_A - c_A)^2}{(\gamma^2 + 2n - 4)^2},$$

$$\Rightarrow \quad 9(\gamma^2 + 2n - 4)^2 > 4(3n - 4)^2.$$

Let us define G as the difference between both sides of this inequation:

$$G = 9(\gamma^2 + 2n - 4)^2 - 4(3n - 4)^2 > 0 \text{ and}$$

$$\frac{\partial G}{\partial n} = 36\gamma^2 - 48 < 0,$$

i.e. G becomes smaller with increasing network strength. Let $\tilde{n}$ denote the maximum level of n which makes G equal to zero:

$$G > 0 \quad for \quad n < \tilde{n} = \frac{20 - 3\gamma^2}{12},$$

which always holds because of $0 < n < 1$ and $0 < \gamma < 1$.

Analogously, we can show that

$$\frac{\partial G}{\partial \gamma} < 0 \text{ and } G > 0 \quad for \quad \gamma < \tilde{\gamma} = \frac{2\sqrt{3}}{3},$$

which always holds due to $0 < \gamma < 1$.

The slope of $CS^e_{A,D}*(c_B)$ is always negative:

$$\frac{\partial CS^e_{A,D}*}{\partial c_B} = -\frac{(1-n)(\alpha_B - c_B)}{\gamma^2} < 0.$$

Since $CS^e_{A,D}*(c_B)$ is a continuous function with a negative slope, $CS^e_{A,D}*(c_B)$ and $CS^e_{A,C}*$ do not intersect within the interval $c_{B,2} < c_B < c_{B,3}$. Thus, $CS^e_{A,C}* > CS^e_{A,D}*$ is fulfilled throughout the deterrence region.

It can be shown in an analogous way that the invitation strategy is welfare-increasing compared to entry-deterrence behavior:

$$W^e_{A,C}* > W^e_{A,D}*(c_{B,2}) \quad \Rightarrow \quad \frac{3(5-3n)(\alpha_A - c_A)^2}{2(3n-4)^2} > \frac{2(3-n-\gamma^2)(\alpha_A - c_A)^2}{(\gamma^2 + 2n - 4)^2}$$

$$\Rightarrow \quad 3(5-3n)(\gamma^2 + 2n - 4)^2 > 4(3-n-\gamma^2)(3n-4)^2.$$

Let us define F as the difference between both sides of this inequation:

$$F = 3(5-3n)(\gamma^2 + 2n - 4)^2 - 4(3-n-\gamma^2)(3n-4)^2 > 0,$$

$$\frac{\partial F}{\partial n} = 36\gamma^2 - 9\gamma^4 - 32 < 0.$$

The upper bound of the network parameter which makes F equal to zero corresponds to:

$$F > 0 \quad for \quad n < \tilde{n} = \frac{12 - 5\gamma^2}{8 - 3\gamma^2},$$

which always holds because of $0 < n < 1$ and $0 < \gamma < 1$.

Analogously, we can show that

$$\frac{\partial F}{\partial \gamma} < 0 \text{ and } F > 0 \quad for \quad \gamma < \tilde{\gamma} = \frac{2\sqrt{3}}{3},$$

which always holds due to $0 < \gamma < 1$.

The slope of $W_{A,D}^{e} * (c_B)$ is negative for $\Pi_{A,D}^{e} * > 0$, as it can be easily seen by solving (4.19) for c_B:

$$\frac{\partial W_{A,D}^{e} *}{\partial c_B} = -\frac{\gamma(\alpha_A - c_A) - (1-n)(\alpha_B - c_B)}{\gamma^2} < 0.$$

Since $W_{A,D}^{e} * (c_B)$ is a continuous function with a negative slope, $W_{A,D}^{e} * (c_B)$ and $W_{A,C}^{e} *$ do not intersect in the interval $c_{B,2} < c_B < c_{B,3}$. Thus, $W_{A,C}^{e} * > W_{A,D}^{e} *$ holds throughout the deterrence region.   *QED*

References

Adams, M. (1996), "Norms, Standards, Rights", M.J. Holler and J.-F. Thisse (eds.), The Economics of Standardization, Special Issue of the European Journal of Political Economy, 12, 363-375.

Akerlof, G.A. (1970), "The Market for 'Lemons': Quality Uncertainty and the Market Mechanism", *Quarterly Journal of Economics*, 84, 488-500.

Arthur, W.B. (1989), "Competing Technologies, Increasing Returns, and Lock-In by Historical Events", *Economic Journal*, 99, 116-131.

Becker, G.S. (1964), *Human Capital*, New York, Columbia.

Beggs, A. and P. Klemperer (1992), "Multi-Period Competition with Switching Costs", *Econometrica*, 60, 651-666.

Belleflamme, P. (1998), "Adoption of Network Technologies in Oligopolies", *International Journal of Industrial Organization*, 16, 415-444.

Belleflamme, P. (2002), "Coordination on Formal vs. De Facto Standards. A Dynamic Approach", *European Journal of Political Economy*, 18, 153-176.

Belman, D. and J. Heywood (1997), "Sheepskin Effects by Cohort: Implications of Job Matching in a Signaling Model", *Oxford Economic Papers*, 49, 623-637.

Besen, S.M. and J. Farrell (1994), "Choosing How to Compete: Strategies and Tactics in Standardization", *Journal of Economic Perspectives*, 8, 117-131.

Blankart, C.B. and G. Knieps (1993), "State and Standards", *Public Choice*, 77, 39-52.

Blankart, C.B. and G. Knieps (1994), "Kommunikationsgüter ökonomisch betrachtet", M. Tietzel (ed.), Ökonomik der Standardisierung. Special Issue of *Homo Oeconomicus*, 11, 449-463.

Choi, J.P. (1996): "Standardization and Experimentation: Ex Ante vs. Ex Post Standardization", M.J. Holler and J.-F. Thisse (eds.), *The Economics of Standardization, Special Issue of the European Journal of Political Economy*, 12, 273-290.

Choi, J.P. and M. Thum (1998), "Market Structure and the Timing of Technology Adoption with Network Externalities", *European Economic Review*, 42, 225-244.

Chou, C. and O. Shy (1990), "Network Effects without Network Externalities", *International Journal of Industrial Organization*, 8, 259-270.

Chou, C. and O. Shy (1996), "Do Consumers Gain or Lose when More People Buy the Same Brand?", M.J. Holler and J.-F. Thisse (eds.), *The Economics of Standardization, Special Issue of the European Journal of Political Economy*, 12, 309-330.

Church, J. and N. Gandal (1992), "Network Effects, Software Provision, and Standardization", *Journal of Industrial Economics*, 40, 85-103.

Church, J. and N. Gandal (1993), "Complementary Network Externalities and Technological Adoption", *International Journal of Industrial Organization*, 11, 239-260.

Church, J. and N. Gandal (1996), "Strategic Entry Deterrence: Complementary Products as Installed Base", M.J. Holler and J.-F. Thisse (eds.), *The Economics of Standardization, Special Issue of the European Journal of Political Economy*, 12, 331-354.

Church, J. and I. King (1993), "Bilingualism and Network Externalities", *Canadian Journal of Economics*, 26, 337-345.

Cooper, R., D.V. DeJong, R. Forsythe and T.V. Ross (1992), "Communication in Coordination Games", *Quarterly Journal of Economics*, 107, 739-771.

Costrell, R.M. (1994), "A Simple Model of Educational Standards", *American Economic Review*, 84, 956-971.

David, P.A. (1985), "Clio and the Economics of QWERTY", *American Economic Review*, 75, 332-337.

David, P.A. (1987), "Some New Standards for the Economics of Standardization in the Information Age", P. Dasgupta and P.L. Stoneman (eds.),
Economic Policy and Technological Performance, Cambridge: Cambridge University Press, 206-239.

de Palma, A. and L. Leruth (1996), "Variable Willingness to Pay for Network
Externalities with Strategic Standardization Decisions", M.J. Holler and
J.-F. Thisse (eds.), *The Economics of Standardization, Special Issue of
the European Journal of Political Economy*, 12, 235-251.

Dixit, A. (1979), "A Model of Duopoly Suggesting a Theory of Entry Barri-
ers", *Bell Journal of Economics*, 10, 20-32.

Dixit, A. (1980), "The Role of Investment in Entry-Deterrence", *Economic
Journal*, 90, 95-106.

Economides, N. (1989), "Desirability of Compatibility in the Absence of Network Externalities ", *American Economic Review*, 79, 1165-1181.

Economides, N. (1996a), "Network Externalities, Complementarities, and
Invitations to Enter", M.J. Holler and J.-F. Thisse (eds.), *The Economics
of Standardization, Special Issue of the European Journal of Political
Economy*, 12, 211-233.

Economides, N. (1996b), "The Economics of Networks", *International Journal of Industrial Organization*, 14, 673-699.

Farrell, J. (1987), "Cheap Talk, Coordination, and Entry", *Rand Journal of
Economics*, 18, 34-39.

Farrell, J. and N.T. Gallini (1988), „Second-Sourcing as a Commitment: Monopoly Incentives to Attract Competition", *Quarterly Journal of Economics*, 103, 673-694.

Farrell, J. and M.L. Katz (1998), "The Effects of Antitrust and Intellectual
Property Law on Compatibility and Innovation", *Antitrust Bulletin*, 43,
609-650.

Farrell, J. and G. Saloner (1985), "Standardization, Compatibility, and Innovation", *Rand Journal of Economics*, 16, 70-83.

Farrell, J. and G. Saloner (1986a), "Installed Base and Compatibility: Innovation, Product Preannouncements, and Predation", *American Economic
Review*, 76, 940-955.

Farrell, J. and G. Saloner (1986b), "Standardization and Variety", *Economic
Letters*, 20, 71-74.

Farrell, J. and G. Saloner (1988), "Coordination through Committees and Markets", *Rand Journal of Economics*, 19, 235-252.

Farrell, J. and G. Saloner (1992), "Converters, Compatibility, and the Control of Interfaces", *Journal of Industrial Economics*, 40, 9-35.

Farrell, J. and C. Shapiro (1988), "Dynamic Competition with Switching Costs ", *Rand Journal of Economics*, 19, 123-137.

Gandal, N., M. Kende and R. Rob (2000), "The Dynamics of Technological Adoption in Hardware/Software Systems: The Case of Compact Disc Players", *Rand Journal of Economics*, 31, 43-61.

Grindley, P. (1995), *Standards Strategy and Policy: Cases and Stories*, New York: Oxford University Press.

Holler, M.J., G. Knieps and E. Niskanen (1997), "Standardization in Transportation Markets: a European Perspective", M.J. Holler and E. Niskanen (eds.), *EURAS Yearbook of Standardization*, 1, Munich: ACCEDO-Verlag, 371-390.

Holler, M.J., B. Layes and R. Winckler (1999), "On Compatibility of Unionized Labor", M.J. Holler and E. Niskanen (eds.), *EURAS Yearbook of Standardization*, 2, Munich: ACCEDO-Verlag, 369-378.

Jeanneret, M-H. and T. Verdier (1996), "Standardization and Protection in a Vertical Differentiation Model", M.J. Holler and J.-F. Thisse (eds.), *The Economics of Standardization, Special Issue of the European Journal of Political Economy*, 12, 253-271.

Jones, P. and J. Hudson (1996), "Standardization and the Costs of Assessing Quality", M.J. Holler and J.-F. Thisse (eds.), *The Economics of Standardization, Special Issue of the European Journal of Political Economy*, 12, 355-361.

Jones, P. and J. Hudson (1997), "The Gains of Standardization from Reduced Search Costs", M.J. Holler and E. Niskanen (eds.), *EURAS Yearbook of Standardization*, 1, Munich: ACCEDO-Verlag, 331-346.

Jørgensen, S. (1997), "Law as a Standardizing System", M.J. Holler and E. Niskanen (eds.), *EURAS Yearbook of Standardization*, 1, Munich: ACCEDO-Verlag, 411-416.

Katz, M.L. and C. Shapiro (1985), "Network Externalities, Competition and Compatibility", *American Economic Review*, 75, 424-440.

Katz, M.L and C. Shapiro (1994), "Systems Competition and Network Effects", *Journal of Economic Perspectives*, 8, 93-115.

Klemperer, P. (1987), "Markets with Consumer Switching Costs", *Quarterly Journal of Economics*, 102, 375-394.

Koski, H.A. (1999), "The Installed Base Effect: Some Empirical Evidence from the Microcomputer Market", *Economics of Innovation and New Technology*, 8, 273-310.

Langenberg, T. (2002), "Signaling und Wohlfahrt", B. Nerré (Ed.), *Forschungsspektrum aktueller Finanzwissenschaft*, Heidenau: PD-Verlag, 9-25.

Langenberg, T. (2005), "Inter-Technology versus Intra-Technology Competition in Network Markets", M.J. Holler (ed.), *EURAS Yearbook of Standardization*, 5, Munich: ACCEDO-Verlag, 21-46.

Layes, B. (1998), *Kompatibilität von Arbeitsqualifikationen*, Frankfurt a.M.: Lang.

Liebowitz, S.J. and S.E. Margolis (1994), "Network Externality: An Uncommon Tragedy", *Journal of Political Perspectives*, 8, 133-150.

Marinoso, B.G. (2001), "Technological Incompatibility, Endogenous Switching Costs and Lock-in", *The Journal of Industrial Economics*, 49, 281-298.

Matutes, C. and P. Regibeau (1988), "Mix and Match: Product Compatibility without Network Externalities", *Rand Journal of Economics*, 19, 221-234.

Matutes, C. and P. Regibeau (1996), "A Selective Review of the Economics of Standardization: Entry deterrence, Technological Progress and International Competition", M.J. Holler and J.-F. Thisse (eds.), *The Economics of Standardization, Special Issue of the European Journal of Political Economy*, 12, 183-209.

Pfähler, W. and H. Wiese (1998), *Unternehmensstrategien im Wettbewerb: Eine spieltheoretische Analyse*, Berlin, Heidelberg: Springer-Verlag.

Riley, J.G. (1976), "Information, Screening and Human Capital.", *American Economic Review*, 66, 254-260.

Saloner, G. and A. Shepard (1995), "Adoption of Technologies with Network Effects: An Empirical Examination of the Adoption of Automated Teller Machines", *Rand Journal of Economics*, 26, 479-501.

Shapiro, C. and H.R. Varian (1999), *Information Rules: a Strategic Guide to the Network Economy*, Boston, Massachusetts: Harvard Business School Press.

Shy, O. (2002), "A Quick-and-Easy Method for Estimating Switching Costs", *International Journal of Industrial Organization*, 20, 71-87.

Singh, N. and X. Vives (1984), "Price and Quantity Competition in a Differentiated Duopoly", *Rand Journal of Economics*, 15, 546-554.

Spence, M. (1973), "Job Market Signaling", *Quarterly Journal of Economics*, 87, 355-374.

Spence, M. (1976), "Product Differentiation and Welfare", *American Economic Review*, 66, 407-414.

Spence, M. (2002), "Signaling in Retrospect and the Informational Structure of Markets", *American Economic Review*, 92, 434-459.

Swann, G.M.P. (2002), "The Functional Form of Network Effects", *Information Economics and Policy*, 14, 417-429.

Thum, M. (1994), "Möglichkeiten und Grenzen staatlicher Standardsetzung", M. Tietzel (ed.), Ökonomik der Standardisierung. Special Issue of *Homo Oeconomicus*, 11, 465-499.

Thum, M. (1995), *Netzwerkeffekte, Standardisierung und staatlicher Regulierungsbedarf*, Tübingen: Mohr.

Wiese, H. (1997), "Compatibility, Business Strategy and Market Structure – a Selective Survey", M.J. Holler and E. Niskanen (eds.), *EURAS Yearbook of Standardization*, 1, Munich: ACCEDO-Verlag, 283-308.

Witt, U. (1997), "'Lock-in' vs. 'Critical Masses' – Industrial Change under Network Externalities", *International Journal of Industrial Organization*, 16, 753-773.

Lecture Notes in Economics and Mathematical Systems

For information about Vols. 1–475
please contact your bookseller or Springer-Verlag